King Arthur in Search of His Dog

and Other Curious Puzzles

Raymond M. Smullyan

DOVER PUBLICATIONS, INC.
Mineola, New York

Copyright

Bibliographical Note

King Arthur in Search of His Dog and Other Curious Puzzles is a new work, first published by Dover Publications, Inc., in 2010.

International Standard Book Number

ISBN-13: 978-0-487-47435-9
ISBN-10: 0-486-47435-6

Manufactured in the United States by Courier Corporation
47435601
www.doverpublications.com

Contents

Guide to the Reader

This volume is indeed for readers of all ages who enjoy puzzles. It is divided into four books, the first of which is mainly for the younger readers and consists primarily of arithmetical puzzles. Some of them are pretty difficult, but if you cannot solve them, you will learn much from reading the solutions, which will show you how you can solve similar problems. Book II is for the young and old alike and consists of logical problems of crime detection. Book III consists of both logical and arithmetical puzzles and should also appeal to young and old alike. Book IV is really my favorite part of this volume, and I believe it will be yours, too! It consists entirely of clever logic puzzles—those that arose in the course of King Arthur's amazing search for his missing dog—and is for *all* readers.

The four books are independent units and can be read in any order. Those who, like me, favor logic puzzles will find them aplenty in Books II and IV (and also in Chapter VII of Book III).

Happy reading!

—Raymond M. Smullyan

BOOK I

DO THESE DOGS PUZZLE YOU?

CHAPTER I
HOW IT ALL STARTED

It all started at Alice's birthday party—not the Alice of Lewis Carroll's *Alice in Wonderland*, but my twelve-year-old friend Alice of my book, *Alice in Puzzleland.* At the party were, of course, lots of children. Now, when I am at a party where there are children, I usually spend much more time with the children than I do with the adults.

"Do you know any puzzles about dogs?" one of them asked.

"Do I know any puzzles about dogs!" I answered indignantly. "Do *I* know any puzzles about dogs! What do you think I am—one who doesn't know any puzzles about dogs?"

"That's just what I'm asking," he replied quietly.

"Of course I know puzzles about dogs!"

"Will you tell us some?" asked another.

"Certainly," I replied.

And so we got started.

•1•

How Much? "Before I tell you a dog puzzle, I began, "I'd like to ask you a question."

"Suppose you and I have the same amount of money. How much must I give you so that you have ten dollars more than I?"

"Why, ten dollars, of course!" said Michael.

"Now, just a minute Mike," I said. "Suppose we each have, say, fifty dollars. If I gave you ten, then you'd have sixty and I would have only forty, hence you would have twenty dollars more than I, not ten!"

"By gosh, you're right!" exclaimed Michael.

So what is the correct answer to the question?

(Solutions to all puzzles are given at the end of each chapter).

•2•

Stray Dogs "Okay, now I'll tell you a puzzle about dogs," I said.

"I once saw some stray dogs walking down a lonely country road. There were two dogs behind a dog, two dogs in front of a dog, and a dog in the middle. What is the smallest number of dogs there could have been?"

"Five," answered Tony, "and they were arranged like this:

"Not so!" said his sister Alice.

Who was right?

•3•

How Did Mary Rescue Her Puppy? Here's a very cute practical one.

Mary's parents had a farm in which there was a square pond. In the middle of the pond there was a small square island. One day Mary's puppy swam to the small island. For some reason, he was afraid to swim back.

Well, Mary looked around for some means to fetch him back. She could not swim, and the water was too deep for her to wade. She found two long planks, but neither was quite long enough to reach from the edge of the pond to the island.

Well, Mary was a very clever and practical little girl, so she figured out a way of placing the planks so she could safely step across, reach the island, and carry the puppy back. She did not nail the planks, nor attach them in any way; they remained quite loose.

How did she do this?

•4•

How Much Did Mary's Puppy Weigh? "I might add," I continued, "that Mary and her puppy together weigh 90 pounds. Mary weighs 60 pounds more than her puppy."

What does the puppy weigh?

•5•

The Bottle and the Cork "This reminds me of a very old puzzle," I said. A bottle and a cork together cost a nickel. The bottle costs four cents more than the cork. How much does the cork cost?"

•6•

A Matter of Direction I once saw two dogs standing in a field facing in opposite directions. One was facing due north and the other was facing south. How can they manage to see each other without turning around, or even turning their heads, or without using any mirrors or other reflecting surfaces?

•7•

How Much Dog Food? "Here's another," I said. "If a dog-and-a-half eats a pound-and-a-half in a day-and-a-half, how much does a dog eat in six days?"

"That's an old one!" said Alice.

"I know," I replied, "but for those who haven't heard it, what's the answer?"

"Six," said Willie.

"Why?" I asked.

"Because, to say that a dog-and-a-half eats a pound-and-a-half in a day-and-a-half, is to say that one dog eats one pound in one day; therefore, in six days, he eats six pounds."

Willie's reasoning was not quite right. Can you find the correct answer?

•8•

Story of the Bookworm "Speaking of old ones," I said, "let me tell you one of the best old ones I know:

"A man had a two-volume dog encyclopedia. The two volumes are resting on a bookshelf in the normal position—Volume I to the left of Volume II. Without the covers, each volume measures one-inch thick.

"A bookworm starts on page 1 of Volume I and wishes to bore his way to the last page of Volume II. How far must he travel?"

•9•

Toby and Dinah "Once upon a time," I began, "a little girl named Alice had two pets named Toby and Dinah."

"What kind of pets were they?" asked one of the group.

"Toby is either a dog or a cat," I answered, "but I won't tell you which. Dinah is also either a dog or a cat, but again I won't tell you which."

"Are they the same or different?" asked another. "I mean, is one of them a cat and the other a dog, or are they both cats or both dogs?"

"It could be that they are the same, or again, it could be that they are different."

"I'll bet that they are not both cats," said Michael.

"I'll bet Toby is a cat," said Timothy.

"Just a minute now," I objected. "What basis do you have for these guesses?"

"None," said Michael, "but were we right or wrong?"

I thought for a moment. "You know," I said, "it's a funny thing! It so happens that either both of you are right, or both of you are wrong, but I won't tell you which!"

From this, the party was able (after a little thought) to deduce whether Toby is a cat or a dog and whether Dinah is a cat or a dog. What are they?

•10•

Another Toby and Dinah "I like that one," said Lillian. "It's a puzzle in pure logic. Can you tell us another like it?"

"Okay," I said. "Another girl, named Betsy, also has two pets, and by a strange coincidence, their names are also Toby and Dinah. Again, each of the pets is either a dog or a cat. Now the following two statements are either both true or both false:

(1) Either Toby is a cat or Dinah is a dog (or both).

(2) Toby is a dog.

What are Toby and Dinah?"

• 11 •

Collie or Greyhound "Good!" said Lillian. "Just give us *one* more like it!"

"All right," I said. "A certain boy has two dogs named Barkus and Wow-wow . . ."

"You can't fool me," cried Tony. "Those are not real names!"

"As a matter of fact, one of them is," I answered. "There was a Roman senator named Barkus. Anyhow, maybe they are both collies, or maybe they are both greyhounds, or maybe one is a greyhound and the other a collie. That is for you to figure out!"

"I'll bet that they are different," said Arthur.

"And I'll bet that Barkus is a collie and Wow-wow is a greyhound," said his brother Bobby.

I thought about this for a minute. "That's very interesting," I said. "If I took you both up on it and bet equal amounts, I would break even."

What is Barkus and what is Wow-wow?

* * *

Story of the Fierce Dog "Speaking of breeds," I said, "do you know the story of the fierce dog? No? Well, a certain man owned an incredibly ferocious dog. Whenever any stranger would come to the house, the dog would simply eat him up! He devoured several hundred people!

"One day, a friend said to the owner: 'My, what a fierce dog! What breed is he anyhow?'

'What breed?' replied the owner. 'I really don't know what breed. All I know is that before I had his nose fixed, he was a crocodile.' "

• 12 •

How Many Dogs and Cats? "How about another arithmetic puzzle?" asked Alice. "I like numbers."

"Only make it about dogs or cats—preferably both," added her younger brother, Tony.

"All right," I replied. "I have just the one for you." I then proceeded to tell them the following puzzle, which led to an extremely interesting conversation that is quite important with regard to many of the later arithmetic puzzles of this book. Here is the problem:

Fifty-six biscuits are to be fed to ten pets, each of which is either a cat or a dog. Each dog is to get six biscuits and each cat is to get five. How many of the pets are dogs and how many are cats?

The first guess was 3 dogs and 7 cats. "That can't be right," I said. "Just think, now, if we have 3 dogs, then, since each dog gets 6 biscuits, the dogs will eat 18 of the biscuits. The 7 cats will eat 35 biscuits (5 apiece), so the total will be 52 biscuits instead of 56!"

The next guess was 1 dog and 9 cats. "Stop guessing," I said, "The idea is to reason it out, not to just stumble on the answer!"

"But," interrupted Alice, "the possible number of dogs is somewhere from 0 to 10, so one could try each of these numbers and see which ones worked and which ones didn't. Wouldn't this constitute a solution?"

"Yes," I replied, "actually, you are right, but such a method of solving the problem is kind of dull and unimaginative. There is a pretty idea behind the problem, and you will never see it that way!"

"The puzzle is real simple for me," said William, an older boy. "I know algebra, and it is easy to solve this problem using algebra."

"True enough," I answered, "but the problem does not require algebra, and there is another way of doing it that is really much neater than the algebraic solution."

Charlie, a small but very clever lad (as you will soon see) looked thoughtful. "You know," he said, "I haven't yet learned algebra; nevertheless, I can see a way of doing this that involves no algebra, and very little arithmetic—just good old-fashioned common sense."

"What is it?" I asked, full of curiosity. Charlie then explained his solution, and all of us were delighted. The method was so wonderfully simple!

Do you have any idea what it was? Look, the matter is quite important, so I'll give you a little hint: Charlie began, "First feed 5 biscuits to each of the 10 pets. Then . . ."

Can you continue? (See solution.)

• 13 •

How Much Does Each Cost? "Do you feel up to a slightly harder puzzle?" I asked.

"Sure," they said bravely.

"Okay, here is an interesting one: A certain pet shop sells cats and dogs; each dog fetches twice the price of each cat. One day a man came in who was exceedingly fond of animals, and he purchased 5 dogs and 3 cats. If, instead, he had bought 3 dogs and 5 cats, he would have spent twenty dollars less. What is the price of each cat and dog?"

"That's too hard!" said Jimmy.

"Well, then, let me give you a little hint: Since each dog is worth 2 cats, how many cats is 5 dogs + 3 cats worth? On the other hand, how many cats is 3 dogs + 5 cats worth?"

• 14 •

Cats and Mice Another pet shop sells cats and pet mice. The proprietor said, "I can give you a cat and 25 mice for forty dollars." The customer replied, "How much would it be if I bought 2 cats and 10 mice?" The proprietor thought for a minute and said, "By a strange coincidence, it would also cost forty dollars."

What is the price of a cat and the price of a mouse?

• 15 •

The Story of Ali and His Pets "Do you know any fairy tales" asked Alice, "maybe like the kind in the *Arabian Nights*?"

"Only, put a puzzle in it, because I love puzzles!" added Nancy.

"All right," I said. "Here is one."

An Arabian boy, Ali, owned some cats and dogs. He had more cats than dogs. One day an evil magician flew over his house and——"

"Just a minute!" interrupted Michael, who is a very practical lad. "I never knew that magicians could fly!"

"Some of them don't," I replied, "but this one did. Anyhow, this evil magician flew over the house and magically transformed one of the cats into a dog. What was Ali's surprise when he awoke next morning and found that he now had the same number of dogs and cats! Well, the next night, a good magician flew over the house and transformed the dog back to a cat. So, when Ali awoke the next morning, things were back to normal. However, on the third night, another evil magician flew over the house, and this time transformed one of Ali's dogs into a cat. When Ali awoke the next morning, he discovered to his amazement that he now had twice as many cats as dogs!"

How many dogs and how many cats did Ali have before all these transformations took place?

A Sequel "I like that story," said Lillian. "Do you have any more like it?"

"Well," I replied, giving myself a mysterious wink in the mirror, "I have a sequel to the story."

"Some time later, Ali sold one of his cats, but he got very little money for it. Why was this?"

"That's silly!" said Alice, after I told them the solution.

* * *

The Story of the Pharmacist "Do you know the joke about the Brooklyn pharmacist and the magic lamp?" I asked. None of them did. "Well," I began, "a Brooklyn pharmacist one day discovered an old oriental-looking lamp in the back of the drug store. He started to rub the dust off, when, lo and behold, sure enough, a Genie appeared! He said, 'You have the good fortune to be the possessor of Aladdin's lamp! I am yours to command. I'll do anything you wish!' The pharmacist heaved a sigh of relief and said, 'Phew, I sure need a rest! I think I'll go out and walk around the town for a couple of hours; meanwhile, you mind the store.'

'Certainly, Master,' the Genie replied.

"Well, it was a slow afternoon, but about half an hour after the pharmacist had left, a woman came into the store and sat at the soda fountain." 'What is your pleasure, Madam?' asked the Genie.

'Oh,' replied the woman, 'make me a vanilla ice cream soda!'

'Surely,' replied the Genie. 'Abracadabra—you are now a vanilla ice cream soda!'"

•17•

The Wisdom of Haroum Al-Raschid "That was even sillier!" exclaimed Alice indignantly. "Come on now, Raymond, tell us a *sensible* one!"

"Okay," I replied with a laugh. "I'll tell you a very old and very good *Arabian Nights* puzzle—it is a serious sequel to the story of Ali.

"When Ali grew up, he went with his friend Ahmed on a pilgrimage from Mecca to Medina. One day, they stopped at a small village for their midday meal. Ahmed had five loaves of bread with him and Ali had only three. Just as they were about to eat, a stranger came up and said that he had no food with him, but plenty of money, and asked if he might share their meal. The two travelers agreed, and the eight loaves were equally divided among the three. After the meal was over, the stranger thanked them, laid down eight coins of equal value, and departed.

"The problem now arose about how the eight coins should be fairly divided. Ahmed proposed that he should take five coins and Ali should take three, since Ahmed had contributed five loaves and Ali three. Ali thought this arrangement unfair; he felt entitled to somewhere between three and four coins, but he admitted to not knowing the exact fraction. Since they could not settle this themselves, they took the problem to the Wali, but the Wali was also unable to settle it.

'Take it to the Kazi,' the Wali suggested. 'He should be able to settle it.' Well, they took the problem to the Kazi.

'Good Heavens,' exclaimed the Kazi, 'even Ebenezer the Magician couldn't solve this one! It must be settled by the Ruler of the Faithful himself!'

"So Haroum Al-Raschid judged the case, surrounded by a throng eager to hear the verdict. To the astonishment of Ali and Ahmed, as well as everyone else present, the Caliph said: 'Let the man who had five loaves take seven of the coins, and the man who had three loaves take only one. Case dismissed!' "

How did Haroum ever get those numbers, 7 and 1?

SOLUTIONS

1. The answer is five dollars. The reason is that whatever amount you and I start with (assuming we have the same amount), if I give you five dollars, then you will have five dollars more than that amount, and I will have five dollars less than that amount, so you will have ten dollars more than I. For example, if we each start with 50 dollars, you will have 55 dollars and I will have 45 dollars.

2. The smallest possible number is not five dogs, but three, arranged like this:

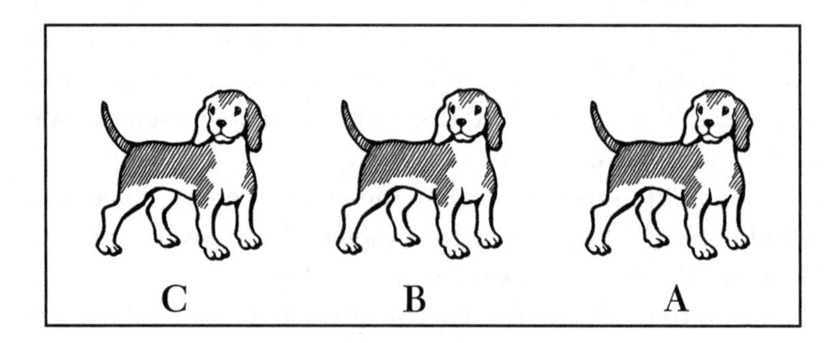

There are two dogs (B and C) in front of a dog (A); two dogs (A and B) behind a dog (C), and a dog (B) in the middle.

3. Mary had the planks arranged like this:

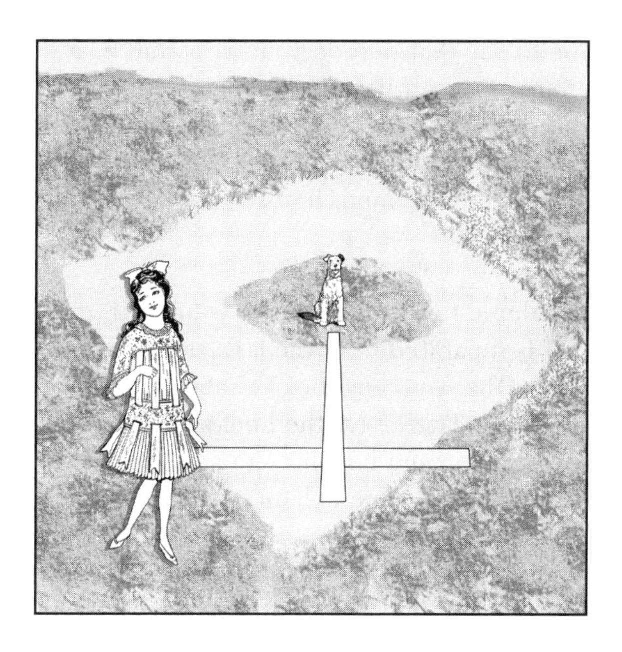

4. A common wrong answer is 30 pounds. Now, if the puppy really weighed 30 pounds, then Mary alone would have weighed 90 pounds (which is 60 pounds more than 30 pounds); hence the two together would have weighed 120 pounds. But the two together actually weighed 90 pounds. So the correct answer is that the puppy weighed 15 pounds and Mary weighed 75 pounds.

5. Again, the usual wrong answer is that the cork cost a penny. But if the cork cost a penny, then the bottle alone would cost 5 cents and the two together would cost 6 cents. But it is given that the two together cost 5 cents. So the correct answer is that the cork cost ½ cent. In other words, the corks sold two for a penny and the bottle cost 4½ cents.

6. The two dogs were facing each other all the time.

7. To say that a dog-and-a-half eats a pound-and-a-half in a day-and-a-half is *not* to say that one dog eats a pound a day—rather, it is to say that one dog eats a pound in a day and a half! (You see, in each day-and-a-half, a dog-and-a-half eats a pound-and-a-half, so *in that same time*, one dog eats one pound.) So, more simply stated, a dog eats one pound in a day-and-a-half, hence two pounds in three days. So a dog eats four pounds in six days.

8. Since Volume I was to the left of Volume II, the first page of Volume I is separated from the last page of Volume II only by the covers (the front cover of Volume I and the back cover of Volume II). Therefore, the bookworm has to bore only through two covers, and not through any pages at all. So the worm has to travel only one-half an inch.

9. It is logically impossible that *both* boys were wrong, because if Michael is wrong, then both pets *are* cats, which would mean that Timothy guessed right that Toby is a cat. Therefore, it is not possible that both guesses were wrong. But I told you that either both guesses were wrong or both were right. Therefore, both guesses were right. Since Timothy's guess was right, then Toby is a cat. Since Michael's guess was also right, they are not both cats—which means that the other pet, Dinah, must be a dog. So Toby is a cat and Dinah is a dog.

10. Again, it is not possible that both statements are false, because if statement (1) is false, then Toby must be a dog and Dinah must be a cat. This would mean that statement (2) is true; hence (1) and (2) can't both be false. But I told you that the statements are either both false or both true. Since they're not both false (which I proved), then they are both true. Since (2) is true, then Toby is a dog. As for Dinah, statement (1) tells us that *at least* one of the following two propositions holds:

(a) Toby is a cat.

(b) Dinah is a dog.

Since proposition (a) does *not* hold (because Toby is a dog, and not a cat), then it must be proposition (b) that holds. Therefore, Dinah is also a dog. So the answer to this problem is that both pets are dogs.

11. To say that if I bet both boys equal amounts I would break even is to say that I would win one bet and lose the other. This means that one of their guesses was right and the other was wrong. Which guess was right and which wrong? Well, if the second guess (Bobby's guess) was right, then the two dogs would be different, so Arthur's guess would also have been right! But it's not true that both guesses were right, so it can't be that Bobby's guess was right. Therefore, it must be that Arthur's guess was right and Bobby's guess was the wrong one. Since Arthur was right, then the two dogs really are different—which means that one of them is a collie and the other a greyhound. Since Bobby was wrong, then it is not the case that Barkus is the collie and Wow-wow the greyhound. So Barkus must be the greyhound, and Wow-wow is the collie.

12. Charlie said, "First feed 5 biscuits to each of the ten pets. Then 6 biscuits are left (50 have just been given to the 10 pets). Now, the cats have already had their portion, so the 6 remaining biscuits must be for the dogs. Also, each dog gets one more biscuit, so there must be 6 dogs. Hence, also, there are 4 cats."

Bravo, Charlie, that was a clever solution! Let us double-check it: Six dogs each eating 6 biscuits totals 36 biscuits for the dogs. Four cats each eating 5 biscuits totals 20 biscuits for the cats. This means 36 + 20, which is 56 biscuits for all ten pets.

13. Since each dog is worth two cats (in the purely monetary sense), then 5 dogs + 3 cats is the same as 13 cats (5 dogs = 10 cats), so 5 dogs + 3 cats = 10 cats + 3 cats = 13 cats). So the man would have paid the same amount had he bought 13 cats instead of 5 dogs + 3 cats. On the other hand, had he bought 3 dogs + 5 cats, it would have been the same as 11 cats (3 dogs = 6 cats, so 3 dogs + 5 cats = 6 cats + 5 cats = 11 cats). So 11 cats cost 20 dollars less than 13 cats. This means that 2 cats cost 20 dollars, and one cat costs 10 dollars. So a cat costs 10 dollars, and a dog costs 20 dollars.

Let us double-check: 5 dogs cost $100 (since 1 dog costs $20), and 3 cats cost $30 (since each cat costs $10), so the man's bill was also $130. Had he bought 3 dogs and 5 cats, he would have paid $60 for the 3 dogs and $50 for the 5 cats, and his bill would have been only $110—which is indeed 20 dollars less than $130.

14. If the man had bought one cat and 25 mice, he would have paid $40. Instead, he took 1 cat more and 15 mice less (in other words, 2 cats and 10 mice, for the same amount). So 1 cat must be worth 15 mice. Therefore, one cat and twenty-five mice is worth the same as 15 mice + 25 mice, which is 40 mice. So 40 mice cost $40, and each mouse costs a dollar, and a cat costs fifteen dollars.

Let us double-check. 1 cat + 25 mice is worth $15 + $25, which is $40. Also, 2 cats + 10 mice is worth $30 + $10 which is also $40.

15. Ali had 7 cats and 5 dogs. After the first magician transformed a cat into a dog, Ali then had 6 cats and 6 dogs (so he had the same number of each). The next day he was back to 7 cats and 5 dogs. Then, the next day, he had 8 cats and 4 dogs, so he then had twice as many cats as dogs.

16. Because it was only an Ali-cat.

17. There are eight loaves to be divided among three people. The easiest way is to cut each loaf into three pieces. We then have twenty-four pieces to be divided among three people, so each person gets 8 pieces. Now, Ali had 5 loaves to begin with, so he had 15 pieces (remember, each piece is one-third of a loaf), and Ahmed had 9 pieces. This means that Ahmed gave only one of his pieces to the stranger, whereas Ali gave 7 of his pieces to the stranger. Therefore, Ali gave seven times as much bread as Ahmed gave, so Haroun Al-Raschid was right!

CHAPTER II
WHAT HAS FOUR LEGS AND BARKS?

Despite Alice's earlier admonitions, the party got sillier and sillier as the afternoon progressed. Perhaps, it was the punch served with the refreshments that made us all a little punch-drunk—I don't know. Anyway, it was a lot of fun, so I'll tell you what happened.

Shortly after the refreshments were served, Mike said to me, "Okay, Mr. Magician, I have one for you. What has four legs and barks?"

"A dog," I answered proudly!

"Oh," said Mike, "I see you've heard that before!"

•1•

My Revenge "Okay," I said to Mike, "I have one for you! What has *five* legs and barks?"

•2•

Then Tony piped up. "I have one," he said. "What is yellow, has four legs, weighs a thousand pounds, and flies?"

•3•

Next, someone asked, "What has four wheels and flies?"

•4•

The next riddle was: "What is orange and squeezes cement?"

•5•

Next, little Charlie (the clever one of the previous chapter) asked, "Divide one hundred by one-half and add seven. What answer do you get?"

Most of them got the wrong answer. What is the correct answer?

•6•

Next, someone asked: "A mathematician bought seven doughnuts. He ate all but three. How many were left?"

•7•

What word, when pronounced wrong, is right, and when pronounced right, is wrong?

•8•

What goes up the chimney down, but does not go down the chimney up?

•9•

At this point, even Alice got into the fun: "I have one for you," she said. "A little girl once said to her mother: 'Mommy, let's you and I go to the railroad station to meet Daddy, and the four of us will come home to dinner.' Now the question is: Why did she say *four* instead of *three*?"

• 10 •

"Here's one," said Mike, "It took a certain man 80 minutes to walk from his house to the railroad station. He took a train to another town to visit his sick aunt. When he came back to the station, he then walked home from the station and it took him an hour and twenty minutes. Yet he walked no slower than before. How do you explain this?"

• 11 •

This reminds me of a beautiful classical problem (whose inventor I do not know). The remarkable thing about this problem is that it is so much simpler than it seems.

A man arrived from work at the railroad station the same time every day—let us say, 6:00 P.M. His wife would arrive at the station with her car exactly at the same time and drive him back home. One day, the man took a train one hour earlier than usual, so he arrived at the station at 5:00 P.M. Since it was a beautiful day, he decided to walk on the road leading to his house to meet his wife on the way. They then drove home and arrived ten minutes earlier than usual.

How long did the man walk before his wife picked him up?

• 12 •

A Follow-up "Why did the man walk part of the way home?" asked Alice. "Why didn't he telephone his wife?"

"Because telephones hadn't been invented yet," I replied.

"That can't be," said Charlie.

Why can't it be?

•13•

"I have one," said Timothy, "and it involves no trick."

"A beggar had a brother, and the brother died; but the dead man had no brother." How do you explain that?

•14•

"I have one," said Charlie. "Which is bigger, six dozen or half a dozen dozen?"

[Mike answered this, but he was wrong!]

•15•

At this point, Tommy, who is the son of the host, showed us a puzzle that delighted us all. He brought out a card that read:

<u>FIND THREE ERRERS</u>
YOU MUST FIND ALL THREE
TO RECIEVE FULL CREDIT

•16•

"I have one," said Barry: "A man went to sleep at 10:00 P.M. after setting the alarm clock for noon the next day. How many hours did he sleep before the alarm woke him?"

•17•

"Here's one," said Lillian: "A wheel has ten spokes. How many spaces are there?"

•18•

"Okay," said Timothy, "what starts when the horse starts?"

• 19 •

"Wise guy," answered Mike. "What goes zub, zub, zub?"

• 20 •

"Wise guy yourself!" replied Timothy, "What goes zzzzzzz?"

• 21 •

"When did Moses sleep five in a bed?" asked Mike.

• 22 •

"I have another for Mike," said Timothy. "What's blue, hangs in the middle of the room, and whistles?"

• 23 •

"Do you want to know something serious?" I asked. "Well, a certain statistician found a definite correlation between the length of one's feet and one's knowledge of mathematics."

"I don't believe it," said Lillian. "You're pulling my leg."

"No, it's true," I replied.

"Then it's just a strange coincidence."

"No, it isn't," I insisted. There really is a perfectly logical explanation for this."

Can you find it?

• 24 •

Next, Paul came out with a cute puzzle: An elderly Arab made a will leaving one-half of his estate to his oldest son, one-third to the middle son, and one-ninth to his youngest son. When he died, all he had in his estate was seventeen camels. Since no one wanted to cut up a camel, everyone was baffled as to what to do. They finally consulted a dervish, who told them what to do.

Can you figure out the solution?

•25•

"Well, boys and girls," I said, "I think it's getting late, and I don't want to overstay my welcome. I thank you all for a delightful afternoon—it was real fun! Meanwhile, let me leave you with an old little puzzle:

"A certain colony of bacteria doubles its volume every minute. If it takes an hour to fill a given container, how long does it take to fill half the container?"

SOLUTIONS

1. Five-fourths of a dog.

2. Two five-hundred-pound canaries.

3. A garbage truck.

4. An orange cement-squeezer.

5. If you divide one hundred by a half, you don't get 50—you get 200! One hundred divided by *two* is 50 (or one hundred *multiplied* by ½ is 50), but one hundred *divided* by ½ is something else.

What does it mean to divide 100 by ½? Well, what does it mean to divide 15 by 3? We say 15 divided by 3 is 5 because we need five 3's to make 15. Also, we need five 4's to make 20, so 20 divided by 4 must be 5. Similarly, to ask what is 100 divided by ½ is to ask how many ½'s we need to make 100. Clearly, the answer is 200. Think of it this way: How many half-dollars do we need to make $100. Fifty half-dollars gives you only $25; we need *200* half-dollars to make $100. Therefore, 100 divided by ½ is 200. If you then add seven, you have 207, which is the correct answer to this problem.

6. Three were left, not four, because I didn't say he ate three. I said he ate all *but* three.

7. The word is "wrong."

8. An umbrella.

9. Because she was too young to count accurately.

10. An hour and twenty minutes is the same as 80 minutes.

11. Let us look at it from the wife's point of view. She arrived home ten minutes earlier than usual; hence, she drove ten minutes less than usual. This means that she drove five minutes less each way. Therefore, in her drive to the station, had she continued normally, she would have reached the station at the usual hour, 6:00 P.M. Hence, she met her husband five minutes before six. But the man had been walking since five o'clock, so he must have walked for 55 minutes.

12. The reason it can't be is that telephones were invented before cars were.

13. The beggar was a woman.

14. The usual wrong answer is that they are the same. But they're not! Half a dozen is the same as six, so half a dozen dozen is the same as six dozen, not six dozen dozen! A dozen dozen is 144, so six dozen dozen is 6 x 144 = 864. But a half a dozen dozen is half of 144, which is 72. So six dozen dozen is 12 times as much as half a dozen dozen.

15. The first error is that the word "ERRERS" is misspelled (it should be "ERRORS"). The second error is that the word "RECEIVE" should have been spelled "RECEIVE." The third error is that there are only two errors, not three! So the word "THREE" should have been "TWO."

If you look deeper, you will see that the above solution is really not valid! Since "THREE" is an error, then there really are three errors after all, so the word "THREE" wasn't really an error. So are there really three errors or only two? There is no correct answer to this question; the whole business is really a paradox—and a funny one at that!

16. With all alarm clocks, except the modern digital ones, the answer would be 2 hours, not 14! In the kind of clock *I* have (and probably you, too), if you set the alarm for 12:00, it will ring on the first 12:00 that comes around, so the man was awakened that same midnight, rather than at noon the next day. With digital clocks, it is different, hence this problem is out-of-date.

17. There are 10 spaces.

18. The horse.

19. A bee flying backwards.

20. A fly, flying either forwards or backwards.

21. When he slept with his forefathers.

22. "A herring," was Timothy's answer.

"But a herring isn't blue," said Mike.

"You can paint it blue," replied Timothy.

"But it doesn't hang in the middle of the room!"

"You can hang it in the middle of the room."

"But it certainly doesn't *whistle*!" screamed Mike.

"I know that! I just said it whistled to make it harder for you to guess."

23. Since many people in the world are children and infants, then, of course, there is a correlation between foot length and mathematical knowledge.

24. Following the advice of the dervish, the three brothers first borrowed a camel from a neighbor. Then they had 18 camels. Well, the oldest brother took half of them, which is 9; the middle brother took one-third of the eighteen, which is 6; and the youngest brother took one-ninth of the eighteen, which is two. Thus, the brothers had 9 + 6 + 2 all told, which is 17. This left one camel, which was returned to the neighbor. This solution, clever as it is, is strictly speaking not valid. Each brother actually got a little more than the will promised. The first brother got not one-half of the estate, which was 17 camels, but one-half of 18 camels. The second brother got one-third of more than the estate, and the youngest got one-ninth of more than the estate—so at any rate, each brother should be satisfied. Actually, the will was a strange one, since $\frac{1}{2} + \frac{1}{3} + \frac{1}{9} = \frac{9}{18} + \frac{6}{18} + \frac{2}{18} = \frac{17}{18}$, so $\frac{1}{18}$ of the estate wasn't willed to anybody!

25. In one hour the container was full. Since the colony doubles its volume every minute, then a minute before the hour was up, the container was half full. So it took 59 minutes to fill half the container.

CHAPTER III
PEEKABOO PUZZLES

Peekaboo is one of my dogs. The puzzles in this chapter for the most part represent true-life situations about Peekaboo and her puppies (now full-grown)—hence I call them "Peekaboo puzzles." They came about in the following manner:

A couple of weeks after the previous party, we had a party at our house, and most of the children whom you already know were present. Some of the grown-ups were inside discussing politics; others were walking together outside (it was a beautiful day), and I was on the porch having fun with the children.

"You have more dogs than that, don't you?" asked Lillian.

"Yes," I said, "some of them are out in the field. I assure you, they'll be back when the refreshments are served—if not sooner!"

"How many dogs do you have?" asked Alice.

That got us started.

•1•

How Many Dogs Do I Have? "It will never do to simply *tell* you how many!" I replied.

"Why won't it do?" asked Alice, who is very logical.

"Because then you won't have the fun of figuring it out for yourself!"

"Look," I continued, "let me tell you this: I have a friend Al who has one dog fewer than I, and I have another friend Jim who has two dogs more than I. It so happens that Al has only half as many dogs as Jim.

How many dogs do I have?

•2•

How Old is Peekaboo? "Is one of these two dogs Peekaboo?" asked Michael.

"No," I replied, "this one is Spotty and that one is Schnible. Peekaboo is the grand lady of the family—she is the mother of all the others. She should be in soon."

"How old is she?" asked Michael.

"Well," I said, "let me put it this way: In another four years, she will be three times as old as she was four years ago."

How old is she now?

•3•

What Color is Peekaboo? "What color is Peekaboo?" asked Michael.

"Well," I replied, "let's play another guessing game. I will tell you that she is either white, or she is black, or she is tan. Why don't you try making some guesses, and when we have enough, I'll tell you some things about the guesses and we will then see if you can deduce her color."

"I'll guess that she is not black," said Michael.

"I'll guess that she is either white or tan," said Tony.

"I'll guess that she is white," said Timothy.

"Hold it!" I said. "We have enough guesses. It so happens that at least one of you made a correct guess and at least one of you made a wrong guess."

What color is Peekaboo?

•4•

How Long is Peekaboo's Snout? At this point the grand lady herself nosed her way through the doggie-door and came into the porch. "Oh, she's pretty," said several of them at once.

"I know," I said proudly.

"She looks just like Schnible," said Tony. "How can one tell them apart?"

"One can't," I kidded.

"Now, Raymond!" said Alice.

"Anyhow," I laughed, "you all see Peekaboo. How long would you say her snout is?"

They all made guesses, and they all came pretty close.

"Why don't we measure it?" Alan suggested.

"Good idea," I said, "but first let's see if you can figure out the length from the following information:

If Peekaboo's snout were three inches longer, it would be twice as long as it would be if it were half an inch less."

How long is Peekaboo's snout?

Postscript—We then all had a lot of fun measuring Peekaboo's snout with a ruler. Peekaboo was wonderful patient—in fact, she wagged her tail and seemed to enjoy it, since all the children were very gentle with her. Of course, our measurements didn't quite agree, since it is difficult to know *just* where the snout begins. However, the average of our readings came quite close to the answer of the problem.

•5•

How Much Did She Cost? "How did you get Peekaboo?" asked Lillian.

"Actually, I got her at a pet shop in New York. It is a rather funny story:

"One day, my wife, Blanche, and I were in New York. 'Why don't you get glasses? You seem to have trouble reading,' said Blanche.

'Good idea,' I replied. 'I'll go to the optometrist on Broadway near 88th Street while you are at the dentist.'

Well, I went to the place, but just next to it is a pet shop, and there in the window I saw this irresistibly cute puppy (then only two months old) lapping milk from a dish.

'That's about the cutest puppy I've ever seen,' I said to myself. 'I bet Blanche would love it!'

I then went in and bought the puppy. Needless to say, in the excitement, I completely forgot all about the optometrist!"

"How much did you pay for her?" one of the children asked.

"Well," I replied, "if I had paid five dollars more, I would have paid five times as much as I would have had I paid fifteen dollars less."

How much did Peekaboo cost?

• 6 •

Enter Snoopy! At this point another dog bungled his way through the doggie-door.

"Ah, here is Snoopy," I said. "Snoopy is my only male dog. He is a little clumsy, and not quite as bright as the others. In fact, I sometimes call him *snupid*."

"That's not nice," said Alice. "I like him. I think he's a nice dog!"

"Of course he's a nice dog!" I answered. "I also love him. Still, sometimes he is a little bit snupid."

"He is more heavy-set and has thicker paws than the others," Tony observed. "How much does he weigh?"

"Well," I replied (remembering the principle of an old puzzle), "he weighs twenty pounds plus half his weight."

"Oh, thirty pounds!" said Michael.

Michael was wrong! How much does Snoopy really weigh?

•7•

And the Other Three? Peekaboo and Schnible together weigh 67 pounds. Schnible and Spotty together weigh 62 pounds. And Peekaboo and Spotty together weigh 65 pounds.

How much does each dog weigh?

•8•

They Ask More About Snoopy "I liked the one about Peekaboo's snout," said Minda. "Can you give us another one about Snoopy's snout?"

"No, give us one about Snoopy's tail," said Minda's older sister, Lenore.

"I'll tell you what," I said. "I'll combine the two, so that in one problem you can figure out the lengths of both his tail and his snout!"

"Well, Snoopy's tail is seven inches longer than his snout. If his tail were four inches shorter, then it would be twice as long as his snout."

How long is the tail, and how long is the snout?

•9•

Abraham Lincoln's Puzzle Abraham Lincoln once proposed the following conundrum: If the tail of a dog were called a leg, how many legs would a dog have?

•10•

A Puzzle-Story About Lincoln As we are on the subject of Abraham Lincoln, let me tell you the following story I once read about him:

Sometime before his presidency, an owner of a railroad company was trying to interest Mr. Lincoln in buying stock in his railroad. After greatly extolling the virtues of his railroad, he said: "What's more, Mr. Lincoln, my railroad system has the added advantage that any collision between trains is impossible." Lincoln thought for a moment and, with his usual logical acumen, replied, "Maybe it is highly *improbable*, but surely not *impossible*! "Yes," roared the owner, "a collision is not merely improbable, it is absolutely *impossible*!"

Can you guess why it was impossible for there to be a collision?

Another Lincoln Story—A book salesman once went to the White House to try to sell President Lincoln a book. Lincoln was not interested. The book salesman pleaded, "Well, Mr. President, since you are not personally interested in this book, couldn't you at least write an endorsement for it to make it easier for me to sell it to others?"

"Certainly," replied Lincoln, who straightaway wrote the following endorsement: "He who likes this kind of book will find it just the kind of book he likes."

•11•

How Many Times Around? "Tell us some more Peekaboo-puzzles," said Lillian.

"All right," I replied, "here is a cute simple one."

One day, Peekaboo entered a barn in the center of which was a large circular haystack. A squirrel was standing in the corner opposite the corner where Peekaboo had entered. When Peekaboo saw the squirrel, she started chasing it around the haystack. It takes the squirrel 40 seconds to run around the stack, but it takes Peekaboo only 30 seconds to make one round of the stack.

How many times around the stack did Peekaboo have to run before she caught the squirrel?

• 11A •

Some Variants (1) Suppose Peekaboo had run a little slower and taken 35 seconds to run around the stack, but the squirrel ran at the same speed as before. How many times around now before Peekaboo caught the squirrel? (2) Suppose Peekaboo takes 30 seconds, but the squirrel takes 35. In which corner of the barn will Peekaboo catch the squirrel—the corner where Peekaboo started, the corner where the squirrel started, or one of the other two corners?

• 12 •

How Fast Does Peekaboo Walk? One day Peekaboo left the house exactly at noon and walked in the direction of the town, which is exactly two miles way. Shortly after, I drove to town to do some shopping. I met Peekaboo just when she arrived in town, and I put her in the car and spent fifteen minutes shopping. I then drove Peekaboo home, and when we reached the house it was seventeen minutes to one. Now, I drive on this road exactly six times as fast as Peekaboo walks.

How fast does Peekaboo walk?

• 13 •

How Fast Does Peekaboo Swim? One day Peekaboo swam a quarter of a mile upstream in 15 minutes. The current was one mile an hour.

How fast does she swim in still water?

• 14 •

What Was the Speed of the Current? On another day, Snoopy swam a quarter of a mile upstream in 7½ minutes. If there had been no current, he would have done it in 5 minutes.

What was the speed of the current?

• 15 •

How Far Did She Go? One day Peekaboo left the house and walked down a road at the rate of 3 miles an hour. After a while, she got a little hungry, and remembering it was close to dinner time, she trotted back to the house twice as fast. She was gone a total of fifteen minutes.

How far did she go?

• 16 •

Spotty Chases a Turtle One day a turtle started down a path at the rate of 10 feet a minute. One hour and twenty minutes later, Spotty, at the top of the path, spied the turtle (which was still crawling at the same rate) and raced after it at the rate of 410 feet a minute.

How long did it take Spotty to catch the turtle?

•17•

The Surprising Generosity of Peekaboo One day Peekaboo and Spotty had biscuits in their bowls; Peekaboo had six biscuits more than Spotty. She generously left some biscuits for Spotty to eat—just enough so that the two dogs had equal amounts.

How many biscuits did Peekaboo leave for Spotty?

•18•

How Many Dog Biscuits? "Here is a much more interesting one," I said. "It is really much less difficult than it seems."

"One day a pile of dog biscuits was lying in a bowl. Snoopy, who is the greediest of my dogs, helped himself first; he ate half of the biscuits and one more. Then Peekaboo came by and ate half of what was left and one more. Then Schnible also ate half of what was then left and one more. Then Spotty ate half of what was left and one more, and the biscuits were all gone.

"How many biscuits were in the bowl, and how many did each dog get?"

SOLUTIONS

1. Some of you will find this a more difficult puzzle than any we have yet considered, so let me tell you in detail the conversation that followed the posing of the puzzle.

"You know," Mike said, "I am having trouble with this. I have tried several numbers, but none of them worked."

"What numbers did you try?" I asked.

"Well, knowing you, I wouldn't be surprised if you had ten dogs, so I first tried the number 10."

"Did it work?" I asked.

"No, it didn't," Mike replied. "If you have 10 dogs, then Al has 9 and Jim has 12. But 9 is not one-half of 12, so that's no good."

"Right," I said.

"Then I tried 5. It also doesn't work—if you have 5, then Al has 4 and Jim has 7, but 5 is not one-half of 7! I could keep trying one number after another, and with luck I would hit on the right one, but isn't there any *logical* way of approaching this problem?"

"Yes there is," I replied. "Let's look at it this way:

"Since Al has only half as many dogs as Jim, Jim has twice as many dogs as Al. So how many more dogs does Al need to have as many as Jim?"

"As many again as he already had," replied Mike.

"Ah!" I replied, "You have hit the vital clue!"

"But Al needs only 3 dogs more to have as many as Jim," said Alice, "because Al has one dog less than you and you have 2 dogs less than Jim. Therefore, Al has 3 dogs less than Jim."

"You're *both* right," I exclaimed, "and that's what gives the answer to this problem! On the one hand, if Al had as many more dogs as he already has, he would have as many as Jim. On the other hand, if he had 3 more dogs, he would have as many as Jim. So 3 must be the same as the number of dogs Al already has. In other words, Al must have 3 dogs."

"So you have 4!" said Tony excitedly.

"And Jim has 6!" said another.

"Right," I said. "I have four dogs.

"Al has one less, so he has three. Jim has two more than I, so he has six. And Al has half as many as Jim, so the answer *four* works."

2. This can be solved using a similar principle. Let's take a dog—call him Rex—that is now four years older than Peekaboo. Let's take another dog—call him Rover—that is now four years younger than Peekaboo. Then Rex's age is the age Peekaboo will be in another four years; Rover's age is the age Peekaboo was four years ago. Then Rex is three times as old as Rover (because Peekaboo's age four years from now is three times what it was four years ago). Now, the difference between three times Rover's age and Rover's age is obviously twice Rover's age. So the difference between Rex's age (which is the same thing as three times Rover's age) and Rover's age is twice Rover's age. But also the difference between Rex's age and Rover's age is eight years (because Rex is four years older than Peekaboo, and Peekaboo is four years older than Rover, so Rex is eight years older than Rover). So, on the one hand, the age difference between Rex and Rover is twice Rover's age, and on the other hand, it is twice Rover's age. This means that eight years is the same thing as twice Rover's age. This means that Rover must be four years old. This makes Peekaboo eight years old and Rex twelve years old. So Peekaboo is eight. Four years ago she was four, and four years from now she will be twelve, which is three times what she was four years ago.

3. If Peekaboo were white, then all three guesses would have been correct. If Peekaboo were black, then all three guesses would have been wrong. But I told you that at least one guess was right and at least one was wrong, so she can't be either black or white. This means she must be tan (and thus the first two guesses were right and the third guess was wrong).

4. Let's take a dog—call her Annie—whose snout is three inches longer than Peekaboo's, and another dog—call her Betsy—whose snout is half an inch less than Peekaboo's. Then Annie's snout is twice as long as Betsy's snout, so the difference between Betsy's and Annie's snout is equal to the length of Betsy's snout. Also, Annie's snout is 3½ inches longer than Betsy's snout, so the difference—3½ inches—is equal to the length of Betsy's snout. So, Betsy's snout is 3½ inches long, which makes Peekaboo's snout 4 inches long (and Annie's snout is 7 inches long). So Peekaboo's snout is 4 inches long. If it were half an inch less, it would be 3½ inches. If it were 3 inches longer, it would be 7 inches, which is twice 3½.

5. The same principle again! Let me put in my right pocket the amount I would have paid had I paid fifteen dollars less, and let me put in my left pocket the amount I would have paid had I paid five dollars more. Then I have five times as much money in my left pocket as I have in my right pocket, so the difference between the amounts in my two pockets is four times the amount in my right pocket. Also, this difference is $20 (because the amount in my left pocket is fifteen dollars more than I paid, and the amount in my right pocket is five dollars less than I paid). So $20 is four times the amount in my right pocket, so I have $5 in my right pocket (and $25 in my left pocket). This means that I paid twenty dollars for Peekaboo.

6. If he really weighed 30 pounds, as Michael said, then half his weight would be 15 pounds—so half this weight plus 20 pounds would be 35 pounds, rather than 30 pounds. So 30 is certainly the wrong answer.

How do we find the right answer? Well, let us look at it this way: Think of Snoopy as divided into two halves (equal halves by weight)—the first half and the second half. Since the first half plus 20 pounds is the whole of Snoopy, then the second half of Snoopy must be 20 pounds. Therefore the entire Snoopy weighs 40 pounds.

Let us check: Snoopy weighs 40 pounds, which is 20 pounds plus half his weight (since half his weight is the other 20 pounds.)

7. Peekaboo and Schnible together weigh 2 pounds more than Peekaboo and Spotty together (since 67 is 2 more than

65). Therefore, Schnible must weigh 2 pounds more than Spotty. Also, Spotty and Peekaboo together weigh 3 pounds more than Spotty and Schnible together (since 65 is 3 more than 62). Therefore, Peekaboo must weigh 3 pounds more than Schnible. So we see that Peekaboo is 3 pounds heavier than Schnible and Schnible is 2 pounds heavier than Spotty.

Now comes the cute part: Imagine Schnible and Spotty together on a scale—the scale registers 62 pounds. Let's take some other dog—call her Sarah—who weighs exactly the same as Spotty. Then Sarah is also 2 pounds lighter than Schnible. Let's take Schnible off the scale and put Sarah on in her place. Then the scale will register 2 pounds less, so it will register 60 pounds. Thus Spotty and Sarah together weigh 60 pounds, which means that Spotty and Sarah each weigh 30 pounds. So Spotty weighs 30 pounds; Schnible, being two pounds heavier, weighs 32 pounds; and Peekaboo, being 3 pounds heavier than Schnible, must weigh 35 pounds.

So Peekaboo's weight is 35 pounds; Schnible's weight is 32 pounds; and Spotty's weight is 30 pounds.

Let us check: Add Peekaboo to Schnible and you get 67; add Schnible to Spotty and you get 62; add Peekaboo to Spotty and you get 65, so everything is hunky-dory!

8. Take a dog—call him Buster—whose tail is 4 inches shorter than Snoopy's tail. Since Snoopy's tail is seven inches longer than his snout, and Buster's tail is 4 inches shorter than Snoopy's, Buster's tail is only 3 inches longer than Snoopy's snout. Also, Buster's tail is twice as long as Snoopy's snout. Thus the difference between Buster's tail and Snoopy's snout is the difference between twice Snoopy's snout and Snoopy's snout—which is one times Snoopy's snout. Therefore, Snoopy's snout is 3 inches and his tail is 10 inches.

Let us check: Snoopy's tail is indeed 7 inches longer than his snout. If the tail were 4 inches less, it would be 6 inches, which is twice the length of Snoopy's snout.

9. Lincoln's answer was *four*. As he wisely explained: "Calling a tail a leg doesn't mean that it is one!"

10. The railroad system had only one train.

11. It takes the squirrel 40 seconds to make one round, so in 30 seconds she makes only three-quarters of a round. Therefore, on every round that Peekaboo makes, the squirrel makes three-quarters of a round, and so Peekaboo has gained one-quarter of a round on the squirrel. Since she has to gain half a round (since they were half a round apart when the chase started), Peekaboo must make two rounds to catch the squirrel.

11A. (1) Since the squirrel takes 40 seconds to make one round, in 5 seconds she makes one-eighth of a round; hence, in 35 seconds she makes seven-eighths of a round. This means that on each of Peekaboo's rounds (which takes 35 seconds), Peekaboo gains one-eighth of a round on the squirrel. Therefore, to gain one-half a round (which Peekaboo must do), she must run around 4 times.

(2) The squirrel now takes 35 seconds to make one round, so she makes one-seventh of a round in 5 seconds. This means she makes six-sevenths of a round in 30 seconds, and 30 seconds is the time of one of Peekaboo's rounds. So in every one of Peekaboo's rounds, one-seventh of a round is gained on the squirrel. Then, after 3½ rounds, Peekaboo will have gained 3½-sevenths of a round—which is the same as half a round, which is what she needs to catch the squirrel. So Peekaboo must run around 3½ times in order to catch the squirrel in the corner where the squirrel started.

12. We reached the house at 12:43 (which is seventeen minutes before one o'clock), so Peekaboo was gone for 43 minutes. Fifteen minutes were spent shopping, so Peekaboo was traveling for 28 minutes—some of the time walking and some of the time in the car. The time walking was 6 times the time driving (because I drive 6 times as fast as Peekaboo walks). This means that we should divide the time—28 minutes—into 7 equal parts, 6 of which she spent walking and one part in the car. Each part is 4 minutes (1⁄7 of 28), so Peekaboo spent 24 minutes walking and 4 minutes in the car (which totals 28 minutes). Therefore, Peekaboo spent 24 minutes walking 2 miles, hence 12 minutes walking 1 mile. So she walked 1 mile in one-fifth of an hour. In other words, she walks 5 miles an hour. This also means that I drive on this road at 30 miles an hour.

Let us check: Peekaboo left at noon and walked 2 miles at the rate of 1 mile in 12 minutes (the same as 5 miles an hour), so she took 24 minutes for 2 miles; hence, when I picked her up in town, it was 12:24. I then shopped for 15 minutes, so when I was finished, it was 12:39. Then I drove back at 30 miles an hour—which is 1 mile in 2 minutes, hence 2 miles in 4 minutes. I spent 4 minutes driving back, so when I reached the house, it was 4 minutes later than 12:39, which is 12:43.

13. To swim a quarter of a mile in a quarter of an hour is to swim a mile in an hour. So she swims against a one-mile-an-hour current at one mile an hour. This means she swims in still water at the rate of two miles an hour.

14. In still water, Snoopy swims a quarter of a mile in 5 minutes, which is one mile in twenty minutes, or 3 miles an hour. But swimming against the current, he swims a quarter of a mile in 7½ minutes, which is one mile in 30 minutes, or 2 miles an hour. The difference between 3 miles an hour and 2 miles an hour was due to the current, which slowed him down 1 mile an hour. Therefore, the current is 1 mile an hour.

Incidentally, it is no coincidence that in the last problem the current was also a mile per hour, because Peekaboo and Snoopy swam up the same stream!

15. Since Peekaboo ran back twice as fast as she went, she went twice as long going as returning. This means that she spent 10 minutes going and 5 minutes returning. So she went away from the house at 3 miles an hour, which is one mile in twenty minutes, but she walked only ten minutes, so she walked away half a mile.

16. An hour and twenty minutes is 80 minutes, so when Spotty spied the turtle, it had already gone 800 feet. Therefore, at the beginning of the chase, the distance between Spotty and the turtle was 800 feet. Now, Spotty gains 400 feet on the turtle every minute (because in one minute the turtle goes 10 feet and Spotty goes 410 feet), so it takes her 2 minutes to gain 800 feet on the turtle.

You can check it this way: Two minutes after the chase started, the turtle was 820 feet from the house. Also, in two minutes Spotty runs 820 feet.

17. This is really the same principle as the first puzzle in this book: If you and I have the same amount of money, how much must I give you so that you have $10 more than I? We recall the answer is $5. Now, let us think of the situation in reverse: Supposing, to begin with, you have $10 more than I. How much must you give me so that we have equal amounts? The answer is also *five dollars*. If you had $6 more than I, instead of $10 more, how much would you have to give me to equalize the amounts? *Answer*: $3.

Well, Peekaboo had six biscuits more than Spotty. To equalize the amounts, Peekaboo gave three biscuits to Spotty.

18. We must solve this problem in reverse order and first ask how many biscuits Spotty found and ate. Well, she ate half of what she found, and one more. Hence, the "one more" was the other half. The other half of what? Obviously, of two; therefore, Spotty found and ate two biscuits. As for Schnible, after eating half of what she found, there was one more biscuit to eat before leaving the two for Spotty. In other words, there were then three. Since three is half of six, Schnible found six biscuits (and ate four, leaving two for Spotty). As for Peekaboo, after eating half of what she found, she left one more than six, which is seven; hence Peekaboo found fourteen biscuits (and ate eight, leaving six for Schnible). As for Snoopy, after eating half of the biscuits originally in the bowl, there was one more than the fourteen that Peekaboo found; thus, there were then fifteen. And so, there were twice fifteen originally in the bowl, which is 30. Thus, the answer is 30.

CHAPTER IV
SHAGGY DOGS

By this time, the dogs had long left the porch and were out in the field playing again.

"Refreshments! Refreshments!" called Blanche from the next room.

A few seconds later, all four dogs, one after the other, tumbled in through the doggie-door.

"I have a theory that these dogs know the word *refreshments*," I said.

"No, you *don't* say!" replied Alice.

Now, this time the refreshments did not include any punch. Rather, there was lemonade served with delicious homemade oatmeal cookies made with real butter! As I say, there was no punch; nevertheless, after the refreshments, we again all got kind of silly. This destroyed my earlier theory that the punch was the cause of our silliness.

First, I suggested that we play the following game with a box of animal-crackers: Two people draw one animal-cracker each from the box, and the one who draws the stronger animal eats them both! It's really a great game, but the only trouble is that ties are fairly frequent. For example, if one draws out a lion and the other a sheep, there is no question who eats them both. But if one draws a hippopotamus and the other a rhinoceros, who wins? Well, I've found from experience that in such cases, it is usually not the stronger animal, but the stronger *person* who wins!

The First Story "Does anyone know any shaggy dog jokes?" asked Tony.

"I do," said Michael. "A man went to see a psychiatrist with the complaint that he often believed he was a dog.

'Really?' said the psychiatrist, 'When did this first start?'

'Oh, ever since I was a puppy.'"

* * *

This reminded me of the story of the man who went to a doctor with the complaint that he couldn't remember things as well as he used to.

"How long has this been going on?" inquired the doctor.

The man looked at the doctor blankly and asked, "How long has *what* been going on?"

* * *

"You probably all know the one about the dog who played chess," I said. "No? Not all of you? Well, a man once saw another man sitting on a doorstep playing chess with his dog.

'What a fantastic dog!' said the man.

'Ah, he's not so good', replied the master. 'I beat him two games out of three.'"

* * *

A theatrical agent once got a telephone call. A voice at the other end said, "I have a great act!"

"What do you do?" inquired the agent.

"I talk."

"What's so remarkable about that?"

"You don't understand; I'm a horse."

I was also reminded of the story of a man who took his dog by the sea. The dog suddenly *walked* way out into the ocean without sinking one bit; he walked around on the water, and then walked back to shore.

The man couldn't believe his eyes! Several times he got the dog to do it again.

Next day, he took the dog out again and brought a friend along—hoping that the dog could repeat this miracle and astound his friend. Sure enough, when they got to the water, the dog easily walked out on the waves and started walking around on the ocean.

"Oh," said the friend, "I see your dog can't swim!"

•1•

The Magic Number Five One of my favorite stories is about the man at the dog race who played hunches. Well, since it was the fifth day of the fifth month of the year, and the fifth race was to start at five minutes past five, and since his fifth wife had just had her fifth baby, which weighed five pounds and five ounces, and since he had exactly five-hundred and fifty-five dollars and fifty-five cents, his hunch told him he should bet on dog #5. He did this. Can you guess what happened?

* * *

"Why are these called *Shaggy-Dog Stories*?" asked Lillian.

"That's hard to say," I replied. "Ever since the original story about the shaggy dog, a lot of dog stories have been called *shaggy-dog stories*."

"Do you know any shaggy-dog *puzzles*?" asked Mike.

"Good idea!" I exclaimed.

•2•

Which Dog Is the Shaggiest? "A man has three shaggy dogs named *Arkus, Barkus* and *Warkus*."

"What funny names!" said Tony.

"I know," I replied. "Shaggy dogs sometimes have shaggy names. Anyhow, the following two facts are true:

(1) Either Barkus or Warkus is the shaggiest.

(2) Either Arkus is the shaggiest or Barkus is the least shaggy."

Which dog is the shaggiest, and which dog is the least shaggy?

•3•

How Much? "Here is another simple one," I began. "A man went into a shaggy-dog shop . . ."

What on earth is a shaggy-dog shop?" asked Alice.

"A pet shop that specializes in shaggy dogs," I replied. "Anyhow, the man went into the shop and took a liking to one particular large shaggy dog.

'What does he cost?' he asked the proprietor.

'You can have him for twenty-two dollars,' was the reply.

The man looked doubtful.

'That includes this beautiful brand-new collar,' added the proprietor, anxious to make a sale.

The man looked around and spied an old shaggy collar in one corner of the shop.

'You know,' he said, 'I think that old shaggy collar would suit him better. How much if I took him with the shaggy collar instead of the new one?'

'With the shaggy collar, you can have him for twenty.'

"Now, it so happens that the shaggy collar was worth one-half as much as the new collar.

"How much would the shaggy dog have cost without any collar?"

•4•

How Many of Each Were Bought? "That was easy!" said little Tony. "Got another?"

"All right. A pet shop one day bought some shaggy dogs at three for forty dollars, and the same number of ordinary dogs at ten dollars apiece . . ."

"Exactly what do you mean by an *ordinary* dog?" asked Alice, who is always very precise.

"For purposes of this problem, let us define an ordinary dog as a dog who is not shaggy," I replied.

"Anyway, the total bill was $140. How many dogs were bought?"

•5•

When Did Rex Pick Up The Newspaper? A friend of mine has a shaggy dog named Rex. Rex, though shaggy, is very intelligent. Every morning Rex is sent out to the village to pick up the newspaper at the stationery store. He then carries the paper back in his snout.

Well, Rex set out one morning, and when he passed the grocery store, it was half-past eight, and Rex had walked one-quarter of the way. When he passed the radio shop, it was twenty-five minutes to nine, and he had walked one-third of the way.

At what time did Rex reach the stationery store?

•6•

Another Shaggy-Dog Shop Another shaggy-dog shop is owned by Mr. McGregor. He has one special shaggy dog for sale named *Shagg*. He is sold along with his brand-new collar for $30. You can buy them separately, but the dog sells for twenty-six dollars more than the collar. How much is the collar alone worth?

•7•

What Is the Original Price? Suppose you buy a shaggy puppy at a 20% discount and pay ten dollars? What is the original price?

•8•

The Shaggy-Dog Sale "This week all shaggy dogs are sold at a twenty-five-percent discount," said Mr. McSnuff proudly, "but next week, you will get only a ten-percent discount."

Well, William procrastinated too long and bought his dog during the second week. If he had bought it the first week, he would have saved three dollars.

How much did William pay?

•9•

How Many Did She Catch? A house was infested with mice, so the owner decided to bring in a shaggy cat.

"I never heard of shaggy cats!" said Alice.

"This one was," I replied, "but she was a good mouser. On the first day she caught one-third of the mice. On the next day, she caught one-third of the remaining mice. On the third day, she caught one-third of the remaining mice. On the fourth day, she caught the remaining eight mice."

How many mice were in the house?

•10•

The Cats Band Together One day the cats of a large deserted house got together and decided to get rid of all the mice. After some time, they did this. We are given the following facts:

(1) Each cat caught the same number of mice.

(2) The number of mice was 143.

(3) The number of mice caught by any one cat was greater than the number of cats.

(4) There were at least two cats.

How many cats were there?

•11•

Annie and Rover To return to our shaggy dogs, there was a shaggy dog, Annie, that had a shaggy puppy named Rover. Annie was born on March 1, 1973. Rover was born on July 1, 1977.

On what date will Annie be three times as old as her puppy?

•12•

Shaggy-Dog Food "A train arrived in the station loaded with cases of shaggy-dog food," I began.

"Now just a minute," said Alice. "What is shaggy-dog food?"

"Food that is fed to shaggy dogs," I replied.

"How does it differ from ordinary dog food?" asked Alice.

"Ordinary dog food is fed to ordinary dogs, whereas shaggy-dog food is fed to shaggy dogs," I answered.

"But is the food itself any different?"

"Not particularly," I replied.

"Then why call it *shaggy*-dog food; why not simply call it *dog food*?"

"You can, if you like," I replied.

Alice was not quite satisfied; she continued to look puzzled.

"Anyway, whatever you call it," I continued, "the important thing is that the train arrived with these cases. The cases had to be transported by trucks to the center of town. There were twenty trucks, and each truck had to make three trips. If each truck had been large enough to hold fifty cases more, then the trucks would each have had to make only one trip.

How many cases were there?

• 13 •

How Many Dogs Did Charles Bring? Alfred, who lived in the mountains, had nine dogs and enough dog food to last them five days. The next day, his friend Charles arrived for an unexpected weekend visit, bringing his own dogs. The remaining food lasted all the dogs for three more days.

Assuming that each dog ate the same amount, how many dogs did Charles bring?

• 14 •

Which Dog Got There First? Two dogs, Edward and Arnold, live six miles apart. They know each other quite well and often meet in Snifftown, which is midway between their homes. One day they left their houses at the same time and went to meet each other in Snifftown. Edward, who is the shaggier, also has the more irregular route: 1 mile is uphill, 1 mile is downhill, and 1 mile is on level ground. Edward walks on level ground at the rate of 4 miles an hour. Uphill, he does only 2 miles an hour, and downhill he can run at 6 miles an hour. Arnold, on the other hand, has a perfectly level route, and he walks steadily at the rate of 4 miles an hour.

Will the two dogs arrive in Snifftown at the same time, or will one get there before the other?

* * *

"Come on, children, it's getting late," said one of the parents. "We've been here a long time and we should get going!"

"No, I want another puzzle!" said Tony. "In fact, I want to hear another Peekaboo puzzle."

"All right, I'll tell you what I'll do," I replied (trying to please both the children and the parents). "I'll give you a hard puzzle to take home with you. Next time we get together, we can discuss the solution, if you haven't already solved it."

•15•

How Long Are the Paths? "You see the beginning of that stream out there," I pointed out to the children, "and you see those two paths leading down to the stream? Well, one path is one-quarter of a mile longer than the other. One day Peekaboo went down the shorter path at the rate of four miles an hour. She returned via the longer path, but she returned at the rate of five miles an hour. Returning took her three minutes less than going."

How long is each path?

SOLUTIONS

1. The dog came in fifth.

2. Since Barkus or Warkus is the shaggiest, then Arkus is surely not the shaggiest. Now, by (2), one of the following two alternatives holds: (a) Arkus is the shaggiest; (b) Barkus is the least shaggy. We have seen that alternative (a) is false; therefore, alternative (b) must be true. In other words, Barkus is the least shaggy. Since the shaggier dog is either Barkus or Warkus, and Barkus is not the shaggiest (in fact, he is the least shaggy), then Warkus must be the shaggiest.

3. The difference in price between the two collars must be $2, and, also, the shaggy collar is half the price of the new collar. Therefore, the shaggy collar is worth $2 and the new collar is worth $4. The shaggy dog cost $18. [With the shaggy collar, the price is $20, and with the new collar, the price is $22.]

4. Three shaggy dogs cost 40 dollars, and three ordinary dogs cost 30 dollars, so six dogs (equally mixed) cost 70 dollars. Since the shop paid 140 dollars, it must have purchased 12 dogs.

Let us check: The shop paid 80 dollars for 6 shaggy dogs and 60 dollars for 6 ordinary dogs, which total $140.

5. By 8:30, Rex has walked one-fourth of the way, and in another 5 minutes he has walked one-third of the way. The difference between one-third and one-fourth is one-twelfth (because one-third is four-twelfths and one-fourth is three-twelfths, and the difference between four-twelfths and three-twelfths is one-twelfth). Therefore, Rex walks one-twelfth of the way in 5 minutes, and he walks the entire way in one hour—also three-quarters of the way in 45 minutes. At 8:30 he still has three-quarters of the way to go, so he must walk another 45 minutes. Therefore, he reaches the stationery store at 9:15.

6. Many people make the same mistake in this problem as in Problems 4 and 5 of Chapter I. The correct answer now is not four dollars, because, if it were, the dog alone would cost $30, and the two together would cost $34 instead of $30.

The correct answer is that the collar costs $2 and the dog costs $28 (which is $26 more than the collar), and the two together, therefore, cost $30.

7. Many people get this problem wrong; they figure that because 20% of ten dollars is two dollars, the original price must have been $12. But this is not right; if the original price had been $12, you would have gotten 20% off of 12 dollars, which is $2.40, so you would have paid $9.60 and no $10! It is instructive to see just where the error lies in the reasoning that led to the conclusion that the original price was $12. The error is this: The difference between the original price and the sale price is 20% *of the original price*—not 20% of the sale price! Those who got the answer of $12 made the mistake of adding 20% of the sale price, rather than 20% of the original price! Now, you might as well ask: Since you don't know the original price, how can you find 20% of the original price? The answer is that we must proceed cleverly! We do the following:

Since you get 20% off, then you are paying 80%, which is $\frac{4}{5}$ of the original price. So the sale price is $\frac{4}{5}$ of the original price. In other words, the sale price times $\frac{4}{5}$ is the original price. This means if you *divide* the sale price by $\frac{4}{5}$, you'll get the original price. Now, to divide by $\frac{4}{5}$ is to multiply by $\frac{5}{4}$. So the original price is $\frac{5}{4}$ of the sale price. And $\frac{5}{4}$ of $10 is $12.50. So the original price was $12.50.

Let us check: The original price was $12.50. Twenty percent of $12.50 is $2.50, so you get $2.50 off and pay $10.

8. First, let's find the original price. William lost 15% of the original price by his procrastination, and he also lost $3, so three dollars is 15% of the original price. Now 15% is 15/100, which is 3/20, so three dollars is 3/20 of the original price, and the original price is 20/3 of $3, which is $20.

William got a 10% discount, so he got $2 off, so he paid $18. This answers the problem. [Note that if he had bought the dog a week earlier he would have gotten 25% off, and 25% of $20 is one-quarter of $20, which is $5. Therefore, in the first week he would have paid $15, which is indeed three dollars less than $18, so the answer checks out.]

9. On the third day, the cat caught ⅓ of the remaining mice and left ⅔ of the mice, which was 8 mice. This means there were 1½ times as many mice on the third day as on the fourth. So, there were 12 mice at the beginning of the third day (and she caught one-third of them, which is 4, and left 8). Likewise, there were 1½ times as many mice on the beginning of the second day as in the beginning of the third—in other words, 18 mice. On the beginning of the first day, there were 1½ times 18 mice—which is 27 mice. So, the answer is 27 mice. On the first day, the cat caught 9 mice (which is one-third of 27) and left 18; on the second day, she caught 6, leaving 12; and on the third day, she caught 4, leaving 8—which she caught on the last day.

Here is another way of solving the problem, which some might prefer (it is how *I* solved it): On the first day, the cat ate ⅓ of the mice, leaving ⅔ of them. On the second day, she ate ⅓ of ⅔, leaving (hariline) ⅔ of ⅔, which is $\frac{4}{9}$. On the third day, she ate ⅓ of the $\frac{4}{9}$, leaving ⅔ of the $\frac{4}{9}$, which is $\frac{8}{27}$, but it is also the 8 mice eaten on the next day. Well, $\frac{8}{27}$ of *what* equals 8? Obviously, $\frac{8}{27}$ of 27 is 8, and so the answer is 27.

10. The only way the number 143 can be expressed as the product of two smaller numbers is $143 = 11 \times 13$. Therefore, either 11 cats caught 13 mice each, or 13 cats caught 11 mice each. The second alternative is ruled out by condition (3).

11. July 1 is 4 months after March 1 (of the same year), so July 1, 1977, is 4 years + 4 months after March 1, 1973. It will be easiest to convert everything into months. Well, 4 years and 4 months is 48 months + 4 months, which is 52 months. So

Annie was 52 months old when Rover was born. At that time, of course, the puppy was zero months old, so Annie was 52 months older than Rover. This means that Annie will always be 52 months older than Rover. [The 52-months difference can never change. If I am 7 years older than you now, won't I be 7 years older than you ten years from now?]

Now, let us consider the state of things on the day that Annie will be three times as old as Rover. At that time, the difference between her age and Rover's age will be twice Rover's age (because Annie will be three times Rover's age), but, also, the difference will be 52 months. Therefore, 52 months is twice Rover's age at that time, which means Rover will be 26 months old at that time, and Annie will be 78 months old. [We see that 78 is both 52 more than 26 and also three times 56.]

So, in 2 years and 2 months (which is 26 months) from July 1, 1977, Annie will be three times as old as Rover. In other words, the date is September 1, 1979.

12. If a truck had carried 50 cases more, it would have saved two trips (it would have made one trip instead of three), which means that if 50 cases had not been added, the truck would have carried 50 cases in two trips, or 25 cases in one trip. Therefore, each truck carried 75 cases in three trips, and, since there were 20 trucks, there were 1,500 cases (75 × 20) all told.

13. Whatever portion of food a dog eats in one day, Alfred has 45 portions to begin with. After the first day, Alfred has 36 portions left. Then Charles's dogs arrive, and the 36 portions lasted all the dogs for three days. Therefore, there were 12 dogs, so Charles brought 3 dogs.

14. Many people arrive at the wrong conclusion that the two dogs will make it to Snifftown in the same time. The error lies in averaging by distance instead of time. Now, if Edward had walked 2 miles an hour, 4 miles an hour, 6 miles an hour for equal periods of *time*, then he would indeed have averaged 4 miles an hour. But, in fact, Edward spent most of his time walking uphill, and the least of his time walking downhill. So his average will be less than 4 miles an hour. It is easy to calculate just how much later he will arrive in Snifftown than Arnold. Arnold walks at 4 miles an hour, which is one mile in fifteen minutes, so to walk three miles, he takes 45 minutes. Now Edward walks one mile at 2 miles an hour, so he takes half an hour to walk uphill. He walks downhill at 6 miles an hour, which is one mile in 10 minutes, so he takes 10 minutes to walk downhill. On level ground he walks one mile in 15 minutes (like Arnold), so he spends 15 minutes on level ground. All together, he walked for 55 minutes (30 + 10 + 15), so he arrived 10 minutes later than Arnold.

15. Four miles an hour is the same as one mile in fifteen minutes; five miles an hour is the same as one mile in twelve minutes. Now, if Peekaboo had come back on the shorter path at the faster rate, she would have saved a quarter of a mile, which at one mile per twelve minutes would have been a further three-minutes saving. So, if she had returned on the shorter path at the faster rate, she would have spent six minutes less than going on the shorter path at the slower rate. Now, the difference between going one mile in fifteen minutes and one mile in twelve minutes is one mile in three minutes. Since the time difference is six minutes, then the distance of the shorter path must be two miles.

Let us check: The short path is 2 miles long, and the longer path is $2\frac{1}{4}$ miles long. Going on the shorter path at the rate of 1 mile in 15 minutes (4 miles an hour) took 30 minutes. Returning on the longer path at the rate of 1 mile in 12 minutes (5 miles an hour) took her 27 minutes (24 minutes for the first 2 miles and 3 minutes for the remaining quarter of a mile), which is 3 minutes less than 30 minutes.

BOOK II

WHO STOLE IT?

CHAPTER V

WHO STOLE WHAT FROM WHOM?

Several weeks later, I got together with the children again at Lillian's house.

"Do you know any of those logical detective puzzles in which one has to figure out who the guilty one is?"

"Why, of course, Lillian," I replied. "They are among my favorites. Let me tell you some."

• 1 •

Who Stole the Dog? A certain dog was stolen one day. Three suspects—Mike, Spike, and Slug—were rounded up for questioning. They made the following statements:

Mike: I didn't steal the dog.

Slug: I stole the dog.

Spike: Slug never stole the dog!

As it happened, at most, one of these three statements was true.

Who stole the dog?

• 2 •

Who Owned the Dog? The dog was recovered. It belonged to one of three boys—Arthur, Bernard, or Charles. They made the following statements:

Arthur: Bernard doesn't own it.
Bernard: That is true.
Charles: Arthur doesn't own it.

As it happened, the real owner was telling the truth, and at least one of the others was lying.

Which boy owns the dog?

•3•

Who Stole the Cat? One day a cat was stolen. Mike, Spike, and Slug were again rounded up for questioning. Mike claimed that Spike had stolen it, and Spike claimed that Slug had stolen it. Now, it was not certain that any of the three suspects had stolen it, but later investigation showed that no guilty person told the truth and no innocent person lied. Also, the cat was not stolen by more than one person.

Can it be determined who stole the cat?

•4•

Who Owns the Cat? The cat belonged to one of three girls—Annabelle, Betsy, or Cynthia. Annabelle claimed that Betsy doesn't own the cat, and Betsy claimed that Cynthia owns the cat. Now, it so happens that the true owner of the cat always tells the truth and is the only one of the three girls who ever tells the truth.

Who owns the cat?

•5•

Who Stole the Horse? One day, a horse was stolen. Again, Mike, Spike, and Slug were rounded up for questioning. This time, each one made two statements. None of them made more than one false statement.

Mike: (1) I did not steal the horse.
(2) The one who stole the horse is Italian.
Slug: (1) Mike never stole the horse.
(2) The one who stole the horse is German.
Spike: (1) I never stole the horse.
(2) It was Slug who stole the horse.

Assuming that one of those three men really stole the horse, which one was it?

•6•

Which Farmer Owned the Horse? The horse was recovered and was to be given back to the rightful owner, who was either Farmer White, Farmer Brown, or Farmer Black. The three farmers each made two statements:

Farmer White: (1) The horse does not belong to Farmer Brown.
(2) It belongs to me.
Farmer Brown: (1) The horse does not belong to Farmer Black.
(2) It belongs to Farmer White.
Farmer Black: (1) The horse does not belong to Farmer White.
(2) It belongs to me.

As it happened, one of the three made two true statements; one made just one true statement; and one made statements that were both false.

Who owns the horse?

•7•

Who Stole the Goat? One day a goat was stolen. Naturally, Mike, Spike, and Slug were the suspects, and, in fact, one and only one of them was guilty. Each of the three accused one of the others, and Mike was the only one who lied.

Was Mike necessarily guilty?

•8•

Who Owns the Goat? The goat belonged to either Farmer White, Farmer Brown, or Farmer Black. Farmer White claimed that the goat was his. Farmer Brown claimed that the goat did belong to Farmer White. Farmer Black either claimed that the goat belonged to him, or he claimed that it belonged to Farmer Brown, but, unfortunately, the court records are confused on this point. At any rate, at least two of the claims were correct.

Who owns the goat?

•9•

Who Stole What? One day, Mike, Spike, and Slug went to the neighboring town of Middleberg and committed three robberies. One of them stole a rifle, one stole some money, and one stole a book. The three were caught, but it was not known which man stole what. At the trial, they made the following statements:

Mike: Slug stole the book.
Spike: Not so; Slug stole the money.
Slug: Those are both lies. I didn't steal either!

As it happened, the one who stole the rifle was lying, and the one who stole the book was telling the truth.

Who stole what?

•10•

Who Stole What From Whom? "And now, we come to a particularly good puzzle," I said to the group proudly.

"Three girls—Abigail, Bernice, and Carol—each had a pet; one was a dog, one a cat, and the other a horse, but we are not told which girl owned which pet. One day, our three villains—Mike, Spike, and Slug—each stole a pet from one of the girls, but it was not known who stole what from whom. The case: proved extremely baffling, but, fortunately, Inspector Craig of Scotland Yard was visiting the country at the time . . ."

"Who is Inspector Craig?" asked Barry.

"He is a character from one of my books," I replied.

"What is the name of this book?" asked Barry.

"You just guessed it!" I said.

"Whatever do you mean?" asked Barry in astonishment.

"I mean just what I said; its name is *What Is the Name of This Book?*"

"Stop kidding us!" said Barry.

"He's not kidding!" said Alice. "I've read the book, and its title really is *What Is the Name of This Book?*, and it really does contain a whole chapter of cases from the files of Inspector Craig."

"Anyway," I intervened, "Inspector Craig was able to find out the following facts, which were enough to solve the case.

(1) The one who stole the horse is a bachelor and is the most dangerous thief of the three.
(2) Abigail is younger than the girl who owns the dog.
(3) Mike's brother-in-law, Slug, who stole from the eldest of the three girls, is less dangerous than the one who stole the dog.
(4) The man who stole from Abigail is an only child.
(5) Mike did not steal from Bernice.

Who stole what from whom?"

SOLUTIONS

1. Since Slug's and Spike's statements contradict each other, one of them must be true. Since only one of the three statements is true, Mike's statement must be false, which means that Mike stole the dog.

Note: It often happens that there are many different ways to solve a given problem. In this problem, for example, another way of solving it is by trial and error: If Mike stole the dog, then Spike's statement is the only true one, so this case is a possibility. If Slug stole the dog, then Mike and Slug both made true statements, so this case is ruled out by the given condition that there is, at most, one true statement. If Spike stole the dog, then Mike and Spike each made a true statement, so this case is ruled out for the same reason. Therefore, the only possibility is that Mike stole the dog.

2. If Charles owned the dog, all three statements would be true, so Charles doesn't own the dog. If Bernard owned the dog, then Arthur's statement would be false; hence Bernard's statement (which agrees with Arthur's) would have to be false, contrary to the given condition that the owner made a true statement. Therefore, Bernard can't be the owner. This means that Arthur must be the owner (and, also, that Charles is the only one who lied).

3. Suppose Mike were guilty. Then Spike is innocent, which means he told the truth, which in turn means that Slug is guilty. But we were told that not more than one person is guilty. Therefore, Mike can't be guilty. Then, since Mike is innocent, he spoke the truth, which means that Spike must be guilty.

4. Suppose Annabelle is not the owner. Then she is lying (since the non-owner is lying), which means that Betsy is the owner. This, in turn, means that Betsy is telling the truth—in other words, that Cynthia must be the owner. Therefore, it is impossible that Annabelle is not the owner. So Annabelle is the owner.

5. Mike's and Slug's second statements can't both be true, so at least one of them is false. Yet Mike and Slug each made at least one true statement. Therefore, either Mike's first statement or Slug's first statement is true. However, these statements say the same thing. Therefore, Mike did not steal the horse. So the thief is either Slug or Spike.

Now, at least one of Spike's two statements is true. If it is the first, then Spike is innocent, which means that Slug is guilty (since we already know that Mike is innocent). If Mike's second statement is true, then, of course, Slug is guilty. So in either case, Slug must be guilty.

6. If Farmer White owns the horse, then he and Farmer Brown would each have made two true statements. If Farmer Brown owns the horse, then he and Farmer Black each made on true and one false statement. Therefore, the horse must belong to Farmer Black (and Farmer Black made two true statements; Farmer Brown made two false statements; and Farmer White made one true and one false statement).

7. Yes! *Reason*: Since Spike accused someone else and spoke the truth, Spike is innocent. Also, Slug accused someone else and spoke the truth, so Slug is also innocent. Therefore, Mike must be the guilty one.

8. If the goat did not belong to Farmer White, then Farmer White's and Farmer Black's claim would both be false, and there would not be two true claims. Therefore, the goat must belong to Farmer White.

9. If Slug stole the rifle, then his statement that he stole neither the book nor the money would be true, but we are given that the one who stole the rifle lied. Therefore, Slug didn't steal the rifle.

If Slug stole the book, then he would have lied in claiming that he stole neither the book nor the money. But we are given that the one who stole the book didn't lie. Therefore, Slug did not steal the book. This means that Slug stole the money.

We now see that Spike told the truth (since Slug did steal the money), which means that Spike didn't steal the rifle. Therefore, he stole the book and Mike stole the rifle.

10. *Step 1*: By (3), Slug is less dangerous than the one who stole the dog and also less dangerous than the one who stole the horse—who is the most dangerous, by (1). Therefore, Slug stole the cat. So Slug, the cat-stealer, is the least dangerous; the dog-stealer is the next most dangerous; and the horse-thief is the most dangerous.

Step 2: By (2), Abigail doesn't own the dog. Also, Slug, who stole the cat, didn't steal it from Abigail, but from the oldest girl (who can't be Abigail), who is younger than at least one of the girls. Hence, Abigail doesn't own the cat. Hence, she owns the horse.

Step 3: Since Abigail owns the horse, the man who stole from her is the horse-thief. Therefore, by (4) the horse-thief is an only child. Also by (1), he is a bachelor. Hence the horse-thief has neither wife nor sister; hence he cannot have a brother-in-law. But Mike has a brother-in-law [by (3)] so Mike is not the horse-thief. He is also not the cat-thief (since Slug is), so he stole the dog. Now we know that Slug stole the cat and Mike stole the dog; hence, Spike stole the horse.

Step 4: Thus, Spike stole from Abigail. Hence, Mike didn't steal from Abigail, and since he didn't steal from Bernice, he must have stolen from Carol. Thus, Mike stole the dog from Carol, Spike stole the horse from Abigail, and Slug stole the cat from Bernice. This settles everything!

CHAPTER VI

THESE STRANGE HUNTERS AND FISHERMEN

"In your last book," said Alice, "you had a lot of interesting puzzles about certain people who always lie, and others who always tell the truth. They make various statements, and the problem is to find out which ones are the liars and which ones are the truthtellers. Do you know any more puzzles like that?"

"No!" said Tony. "I want some more detective puzzles. I want to know who stole it and who owns it!"

Well, at this point, about half the children clamored for more detective puzzles and half clamored for puzzles about people who either always lie or always tell the truth.

"I'll tell you what," I said. "I'll give you some very interesting puzzles that combine both features you want."

And so, I told the following story:

* * * *

"There is a strange tropical island on which hunting and fishing are the only two occupations. Every islander is either a hunter or a fisherman, but never both. The curious thing is that the hunters always lie—they never tell the truth—and the fishermen always tell the truth—they never lie."

"Just a minute," interrupted Alice. "Is this realistic? Is there any evidence that hunters lie any more than fishermen?"

"Of course not," I replied. "This is only a *story*, and I am not saying that in general hunters lie more often than fishermen, but only that on this particular island, it so happened that the hunters always lied and the fishermen always told the truth. Now, I shall tell you some interesting things that happened on this island."

•1•

Who Stole the Monkey? One day a monkey was stolen from the zoo. An islander was tried and was asked, "Did you steal the monkey?" He replied, "The monkey was stolen by a hunter."

Now remember: hunters always lie and fishermen always tell the truth.

Was this islander innocent or guilty?

•2•

What Next? Then it was definitely found out that the thief must be one of two brothers. [Just because they were brothers doesn't mean that they necessarily have the same occupation; it could be that one is a hunter and the other a fishermen]. Well, the two brothers made the following statements in court:

First Brother: Either I am a hunter, or the monkey was stolen by a fisherman.

Second Brother: I didn't steal the monkey!

Who stole the monkey?

•3•

Who Stole the Elephant? One day an elephant (of all things!) was stolen on this island. The two suspects are Aaron and Barab. It was not known at the beginning of the trial whether either one was actually guilty. Well, they made the following statements in court:

Aaron: I did not steal the elephant.

Barab: One of the two of us is a hunter and the other is a fisherman.

Can it be determined who stole the elephant?

•4•

Who Owns the Elephant? Well, the elephant was recovered, and it belonged to one of three islanders. They made the following claims:

First Islander: The elephant is mine.
Second Islander: The elephant is mine.
Third Islander: At least two of us are hunters.

•5•

Who Stole the Panther? On the same island, three men—Alu, Bomba, and Kuhla—made the following statements about a stolen panther:

Alu: Either Bomba is innocent or he is a fisherman.
Bomba: Either I am innocent or Alu is a hunter.
Kuhla: The guilty one is not a fisherman.

Who stole the panther?

•6•

An Intriguing Mystery Finding the owner of the panther proved to be a particularly interesting problem.

"A panther is a rather odd thing to *own!*" said Alice.

"These were odd islanders," I responded. "Anyway, it was known that the panther belonged to one of three men—A, B, or C—though it was not known whether the owner was a hunter or a fisherman. The three made the following statements in court:

A: The panther belongs to C.

B: The panther does not belong to me.

C: At least two of us are hunters.

From this, the judge could not decide who owned the panther. Fortunately, Inspector Craig of Scotland Yard happened to be vacationing on this island at the time and was sitting in court, since he was interested in the case. He asked the judge, 'Your Honor, may I interrogate one of the three?'

'Surely,' the judge replied.

Well, Inspector Craig asked for C, 'Come on, now, which of you three really owns the panther?' C answered, and Craig then knew who owned the panther."

Who owns the panther?

•7•

Did John Steal the Giraffe? On the same island of hunters and fishermen, a giraffe was stolen one day. One of the islanders, John, was accused of stealing it. He and his brother Dick made the following statements in court:

John: I am innocent.

Dick: My brother and I have the same occupation (hunter or fisherman).

Did John steal the giraffe?

•8•

Then Who Did Steal It? Well, *somebody* stole the giraffe. Who was it? After the island police investigated the situation, the thief was narrowed down to one of three islanders—A, B, or C. It was known that only one of the three participated in the robbery. The three made the following statements:

A: B stole the giraffe.

B: The giraffe was stolen by a fisherman.

C: All three of us have the same occupation (hunter or fisherman).

Who stole the giraffe?

• 9 •

Who Stole the Fish? One day a hunter stole a fish from a fisherman. The three suspects were A, B, and C. The hunter who stole the fish was one of them, and the other two were both fishermen:

The judge first asked A, "Did you steal the fish?" A refused to answer. Then B was asked, "Did you steal the fish?" B also refused to answer. Then C was asked, "Did you steal the fish?" C replied, "Either B or I stole the fish."

Who stole the fish?

• 10 •

Who Owns the Fish? The fisherman who owned the fish appeared in court with two others who happened to be both hunters. Call the three men D, E, and F.

D claimed that E owns the fish, and E and F made no comments.

Who owns the fish?

• 11 •

Who Stole the Seal? One day a fisherman stole a seal from another fisherman. The thief was actually one of three suspects:

A, B, or C. It was not known what the others were (hunters or fishermen). They made the following statements:

A: At least one of us is a hunter.

B: I am a fisherman.

C: That is true.

Who stole the seal?

• 12 •

Who Owns the Seal? Three men—A, B, and C—appeared in court. The fisherman who owned the seal was one of them. Only two of them made statements.

A: At least one of us is a hunter.

B: C owns the seal.

Who owns the seal?

• 13 •

The Society of Crafty Hunters A certain subgroup of the hunters have formed a society called "The Society of Crafty Hunters." To be admitted to the society, you must appear before their tribunal and convince them that you are a hunter and that you are crafty. But you are allowed to make only one statement. Assuming you are one of the islanders and are, in fact, a crafty hunter, what statement would gain you entrance into their society?

* * * *

At this point, one of the children objected to the solution given at the end of the chapter for reasons I explain at the end of the solution. So I gave them the following problem to clarify an important logical point.

• 14 •

Some Theoretical Questions

(a) Is it possible for any inhabitant of this island to say, "I am a hunter"?

(b) Is it possible for an inhabitant to say, "I am a hunter and two plus two equals five"?

(c) Is it possible for an inhabitant to make the following two statements separately: (1) I am a hunter; (2) Two plus two equals five?

(d) Can any inhabitant say, "I am a hunter and two plus two equals four?"

• 15 •

The Society of Wise Fisherman To enter the Society of Wise Fishermen, you must make a single statement that will simultaneously accomplish three things:

(1) It must convince them that you are a fisherman.

(2) It must convince them that you have caught at least one hundred fish.

(3) It must enable them to deduce your first name.

What single statement will do this?

• 16 •

How Many Were Married? I once met two islanders, A and B, and I was interested in knowing of each whether he was married or single. Well, A said that B is not an unmarried fisherman, and B said that A is not a married hunter. As I later found out, at least one of them was a hunter.

How many of them were married?

•17•

An Interesting Personal Adventure I was once sent over to the island of hunters and fishermen to do some counter-espionage. [If you don't know the word *espionage*, either ask someone or look it up in a dictionary.] In particular, the government knew that there was a certain man on the island named McSnoy, and it was vital to find out whether McSnoy was a hunter or a fisherman.

Well, shortly after I arrived, I found McSnoy (whom I recognized from a photograph) having lunch with another islander whose name was *McEldridge*. Now, I didn't care in the least whether McEldridge was a hunter or a fisherman; I was interested only in McSnoy. I asked McEldridge, "Are both of you hunters?" McEldridge answered *yes* or *no*. I thought for a while but could not determine what McSnoy was. Then I asked McSnoy, "Did he answer truthfully?" McSnoy answered *yes* or *no*, and I was then able to make an accurate report to the government concerning McSnoy's occupation.

Is McSnoy a hunter or a fisherman?

•18•

Who Owns the Hunting Dog? On the hunter-fisherman island, a hunting dog was lost. It was recovered, and, of course, it belonged to a hunter. The owner was one of two men, A and B, and the other man was a fisherman.

The two men appeared in court. The judge asked A, "If B were asked whether he owns the dog, what would he say?"

A replied, "B would claim to own the dog."

Which one owns the dog?

•19•

Who Stole the Platypus? One day a hunter stole a platypus from the zoo. Three defendants—A, B, and C—were tried. It was not certain that the thief was among them, but it was certain that if any of the three was a hunter, then one of them was the thief. The three made the following statements in court:

A: At least one of us is a fisherman.

B: At least one of us is a hunter.

C: I am not a hunter.

Is the thief necessarily present? If so, can it be determined which one he is?

• 20 •

Who Stole the Aardvark? It was not known whether the thief who stole the aardvark was a hunter or a fisherman. The only suspect was a man named Momba. He was asked in court: "Was the aardvark stolen by a fisherman?" Momba answered the question, and the judge then knew whether he was innocent or guilty.

Which was he?

* * * *

And now, we come to my favorite case of all that happened on this island:

One day, a whale was stolen. Now, please don't ask me *how*, because I haven't the faintest idea! Anyway, a whale *was* stolen, and the problem, of course, is to find out who stole it.

• 21 •

The First Trial (A Case of Identity) It was suspected, but not known, that the thief was one of a pair of identical twins—the only such pair on the island. Now, the twins are not necessarily of the same occupation; it is quite possible that one could be a hunter and the other a fisherman, but then again, they might both be hunters or both fishermen. Anyway, the day of the trial came, and the judge asked them, "Did either of you steal the whale?" He got the following answer:

First Twin: Maybe one of us stole the whale, or then again, maybe neither of us stole the whale.

Second Twin: No fisherman on this island ever steals whales!

Well, this evidence was quite insufficient to convict or acquit either one. The two brothers were returned to their cells.

The next day, the trial resumed, and the twins were brought back to court. The judge suddenly realized that he could not tell them apart; hence, he had no idea which twin had made which statement the day before. He asked one of them—call him A—, "Are you the one who claimed yesterday that no fisherman on this island ever steals whales?" A answered, and the judge then knew whether he was guilty or innocent.

Is A guilty or innocent? Is his brother guilty or innocent?

• 22 •

The Second Trial The next two suspects were Momba and an islander named *Karl*. It is possible that neither of them stole the whale, or that just one of them stole the whale, or that they both stole it together.

Well, eight witnesses testified at this trial. We will call them A, B, C, D, E, F, G, and H. Each of them was either a hunter or a fisherman. They gave the following testimony:

A: Karl is a fisherman.
B: Momba is a hunter.
C: A is a hunter.
D: B is a hunter.
E: C and D are both fishermen.
F: Either Karl is a fisherman, or Momba is a hunter.

G: E and F have the same occupation.
H: My occupation is the same as G's, and at least one of the defendants is innocent.

Out of this logical tangle, the guilt or innocence of each of the two defendants can be determined.

What is the solution?

SOLUTIONS

1. Suppose the speaker is a fisherman. Then his statement is true, which means that the monkey was stolen by a hunter; so the speaker is innocent in this case. On the other hand, suppose the speaker is a hunter. Then his statement is false, which means that the monkey was stolen by a fisherman; so, in this case, the speaker is also innocent. So, in either case, the speaker is innocent.

2. The first brother *claims* that at least one of these alternatives holds: (a) he is a hunter; (b) the monkey was stolen by a fisherman. If he is a hunter, then alternative (a) does hold, which means that his statement is true that at least one of the alternatives holds. This would mean that a hunter made a true statement, which is not possible. Therefore, the first brother can't be a hunter; he must be a fisherman. Now that we know he is a fisherman, then we see that his statement is true, which means that either he is a hunter or else the monkey was stolen by a fisherman. But he is not a hunter; therefore, the monkey was stolen by a fisherman.

Now we know that the first brother is a fisherman, and also that the monkey was stolen by a fisherman. From this information it would be premature to conclude that the first brother must have been the one who stole the monkey; we must reason further.

If the second brother is a hunter, then he didn't steal the monkey, because the monkey was stolen by a fisherman. On the other hand, if the second brother is a fisherman, then his statement is true, which means he, again, couldn't have stolen the monkey. Therefore, it must have been the first brother who stole the monkey.

3. From Barab's statement, we can prove in the following manner that Aaron must be a hunter:

Suppose Barab is a fisherman. Then his statement is true, which means that one of them is a hunter and one a fisherman. Since Barab is a fisherman, then it must be Aaron who is the hunter.

This proves that Aaron is a hunter, provided that Barab is a fisherman. But suppose Barab is a hunter. Well, in that case, his statement is false, which means that Barab and Aaron are not of different occupations (as Barab claims) but must really be of the same occupation. Then, since Barab is a hunter and Aaron is of the same occupation, then Aaron must also be a hunter.

This proves that regardless of whether Barab is a hunter or a fisherman, Aaron must be a hunter. [Incidentally, it is not possible to determine whether Barab is a hunter or a fisherman.] Since Aaron is a hunter, his statement is false, and, therefore, Aaron stole the monkey.

4. The first and second islanders can't both be fishermen, since their statements can't both be true. Therefore, at least one of the first two must be a hunter. Now, if the third islander were a hunter, then it would be true that at least two of them are hunters (namely, he and one of the first two), and we would have a hunter making a true statement. Therefore, the third islander must be a fisherman. This means that his statement is true, so there are at least two hunters present. Since the third islander is not a hunter, then the first two are both hunters. Therefore, both their statements are false, so the third islander must be the owner of the elephant.

5. We first show that Kuhla must be innocent. Well, Kuhla is either a fisherman or a hunter. Suppose he is a fisherman; then his statement that the guilty one is not a fisherman must be true; so Kuhla, since he is a fisherman, can't be guilty. So, if Kuhla is a fisherman, then he is innocent. Now, suppose Kuhla is a hunter; then his statement is false, which means that the thief is a fisherman; so Kuhla, since he is a hunter, can't be the thief. So, in this case, Kuhla is again innocent. Therefore, regardless of whether Kuhla is a fisherman or a hunter, he is innocent. So the guilty one is either Alu or Bomba.

We next prove that Alu is a fisherman—we show that if Alu is a hunter, we get a contradiction. Well, suppose Alu is a hunter; then Bomba must be a fisherman (because he claims that one of the two alternatives holds: (a) Bomba is innocent; (b) Alu is a hunter. Well, alternative (b) does hold; so it is true that at least one of the alternatives (a) or (b) holds). Since Bomba is a fisherman, then it is certainly true that *either* he is innocent *or* he is a fisherman. But then, how could Alu, a hunter, make this true statement? This proves that Alu can't be a hunter; so he is a fisherman.

Since Alu is a fisherman, his statement is true; so Bomba is either innocent or he is a fisherman. So if Bomba is not a fisherman, he is innocent. What if Bomba is a fisherman? In that case his statement is true, which means that either he is innocent or Alu is a hunter. But Alu is not a hunter (we proved that); so Bomba must be innocent. Therefore, Bomba is innocent, regardless of whether he is a fisherman or a hunter.

We now know that Kuhla and Bomba are both innocent; therefore, Alu must be the guilty one.

6. On the basis of the three statements made before Inspector Craig interrogated C, we will show that if C is a hunter, then he owns the panther, and if C is a fisherman, then B owns the panther.

Suppose C is a hunter; then his statement is false; hence there are not at least two hunters; so A and B must both be fishermen. Since A is a fisherman, then his statement is true; so the panther belongs to C. This proves that if C is a hunter, the panther belongs to C.

Suppose C is a fisherman. Then his statement that at least two of them are hunters must be true; hence A and B must both be hunters. Since B is a hunter, his statement is false, which means that B owns the panther. This proves that if C is a fisherman, then B owns the panther.

This is as much as we can deduce from just the three statements made prior to Craig's question to C. Now, Craig asked C who owns the panther, and C either said that he did or that B did or that A did—we are not told which—but we are told that after C answered, Craig did know who owned the panther. Now, C is either a fisherman or a hunter. If C is a fisherman, then B must own the panther (as we have seen); hence C, since he is truthful, would name B as the owner. So, if C is a fisherman, he named B.

Suppose C is a hunter; then, as we have seen, C owns the panther; hence C, since he is untruthful, would never admit owning the panther; hence, he would have named B or C. So if C is a hunter, he named either B or C. So, in either case (fisherman or hunter), C must have named either A or B. If C named B, then it could be either that C is a fisherman and B owns the panther, or that C is a hunter and C owns the panther, but there would be no possible way to know which. Therefore, if C named B, Craig would have had no way of knowing who really owned the panther. However, we are given that Craig *did* know; therefore, it must be that C named A, and Craig then knew that C owns the panther. Therefore, C owns the panther.

7. If Dick is a fisherman, then his statement is true; hence the two brothers are of the same occupation, which means that John is also a fisherman. If Dick is a hunter, his statement is false, which means that John is of a different occupation from Dick, which means that John is a fisherman. In either case, John is a fisherman. Therefore, John's statement is true; so John is innocent.

8. Suppose C is a fisherman. Then all three are fishermen (as he said); hence A is a fisherman, and B stole the giraffe. So if C is a fisherman, then B stole the giraffe.

Now, suppose C is a hunter. Then his statement is false, which means that they don't all have the same occupation; hence, at least one of A or B must be a fisherman. If A is a fisherman, then again B stole the giraffe.

Suppose A is a hunter. Then B is the only fisherman, and his statement is true; so a fisherman stole the giraffe, and since he is the only fisherman, then again B stole the giraffe. So, in all possible cases, B stole the giraffe.

9. Suppose C is a hunter. Then C stole the fish (since C is the only hunter). Then his statement that either B or he stole it is true, and we have the impossible situation of a hunter making a true statement. Therefore, C must be a fisherman. Hence, his statement is true; so either he or B stole the fish. Since C is a fisherman, he didn't steal it. So B stole the fish.

10. If D is a fisherman, then his statement is true that E owns the fish, but this would mean that E is a hunter and owns the fish, which is not possible. Therefore, D must be a hunter. This means that his statement is false; hence, E doesn't really own the fish. Also, D doesn't own it (because D is a hunter); hence, it is F who owns the fish.

11. If A were a hunter, then it would be true that at least one of them is a hunter; so a hunter (A) would be making a true statement, which is not possible. Therefore, A must be a fisherman. Since A is a fisherman, then his statement that at least one of them is a hunter must be true. Therefore, either B or C is a hunter. But since C agrees with B, and at least one of them is a hunter, then both of them must be hunters. Therefore, A is the only fisherman; so A stole the seal.

12. Again, A must be a hunter (for the same reason as in the last problem), and at least one of B or C must be a fisherman. If B is a fisherman, then his statement is true, which means that C owns the seal. If B is a hunter, then he and A are both hunters; hence C is the only fisherman; so C owns the seal. So, in either case (whether B is a fisherman or a hunter), C owns the seal.

13. One statement that would gain you entrance into the society is" "I am a hunter, but not a crafty one." A fisherman could never make that statement, so the tribunal would know that you are a hunter. They would also know that if you were not a crafty hunter, your statement would be true, and that a hunter can't make true statements. Therefore, they will know that you must be a crafty hunter.

Incidentally, some readers may object to this solution on the grounds that no inhabitant of this island can say that he is a hunter. Although it is true that no inhabitant can say that, the objection is not valid for reasons we discuss below.

14. (a) Certainly not! A fisherman would never lie and say that he is a hunter, and a hunter would never truthfully admit to being a hunter. So, no inhabitant of the island can say that he is a hunter.

(b) This is a very different story! No fisherman could say that, but a hunter could. Since it is false that two plus two equals five, then it is false that the speaker is a hunter *and* that two plus two equals five; hence, a hunter could make that false statement.

It is important to realize that if a sentence consists of two clauses connected by the word *and*, if so much as one of the clauses is false, then the whole sentence is to be regarded as false. For example, if I know French but not German, then I would be lying if I said, "I know both French and German," or, if I said, "I know French and I know German."

(c) This is a horse of a different color! No, it is *not* possible for an inhabitant of this island to make these two statements separately, because no one can say (1) alone.

This, in comparison with (b), elucidates a very interesting point in connection with the logic of lying: If a truthful person makes two statements separately, it is the same thing as if he makes one single statement asserting that both are true. But it is an entirely different matter when a liar asserts two statements separately, and when he, in one sentence, asserts that both are true. In the first case, both statements must be false, while in the second case, all that follows is that at least one is false. For example, suppose a liar makes two separate statements: (1) John is guilty; (2) Jim is guilty. Then, in fact, both John and Jim must be innocent. On the other hand, suppose a liar makes the one single statement: John is guilty and Jim is guilty. The liar is in effect saying that both are guilty, and since he is lying, they are not both guilty. But this does not mean that they both have to be innocent; it only means that at least one has to be innocent.

(d) No.

15. Let's say that your first name is John. Then a single statement that will gain your entrance into the Society of Wise Fishermen is this: "Either I am a hunter, or I, whose name is John, have caught at least a hundred fish."

Your statement asserts that at least one of the following two facts holds: (1) you are a hunter; (2) your name is John, and you have caught at least one hundred fish. If you were a hunter, fact (1) would hold, which would make your statement true, but hunters can't make true statements. Therefore (the Society would reason), you must be a fisherman. Then what you say is really the case, which means that either (1) or (2) holds. But (1) can't hold; therefore (2) holds; so your name is John, and you have caught at least one hundred fish.

16. If B is a hunter, then his statement is false, which means that A is a married hunter and, hence, a hunter. If B is not a hunter, then A must again be a hunter, because at least one of them is a hunter. Therefore, A is a hunter. Since A is a hunter, his statement is false; therefore, B is an unmarried fisherman. Since B is a fisherman, his statement is true; so A is not a married hunter. But A is a hunter; therefore, A is unmarried. So, neither one is married.

17. This is quite a different type of problem from any we have yet considered; it appears at first as if you are not given enough information to solve it, but actually you are.

I did not tell you what either McEldridge or McSnoy answered, but it is possible for you to figure out both. Suppose McEldridge had answered, "Yes." Then I would have known what McSnoy was because a fisherman could not claim that he and McSnoy are both hunters; hence McEldridge would have to be a hunter. Therefore, McEldridge's answer was a lie, which means that they are not both hunters; so McSnoy must be a fisherman. So, if McEldridge had answered, "Yes," then I would have known that McSnoy was a fisherman. But, as I told you, I didn't know what McSnoy was; therefore, McEldridge wouldn't have answered, "Yes"; he must have answered, "No." This means that there are three possibilities:

(1) McEldridge is a fisherman and McSnoy a hunter.
(2) McEldridge is a fisherman and McSnoy a fisherman.
(3) Both are hunters.

[The possibility that McEldridge is a hunter and McSnoy a fisherman is out, because in this case McEldridge would have lied and said, "Yes."]

Next, I asked McSnoy whether McEldridge had told the truth. In other words, whether McEldridge is a fisherman. If the first case holds (McEldridge a fisherman and McSnoy a hunter), McSnoy would answer, "No." If the second case holds, he would answer, "Yes," and if the third case holds, he would answer, "Yes." So, if McSnoy had answered, "Yes," then either the second or third case could hold, which means McSnoy could be either a fisherman or a hunter, and I couldn't have known which. But I told you that McSnoy's answer *did* enable me to know which; therefore, McSnoy must have answered, "No," and I knew that McSnoy must be a hunter (and also, incidentally, that McEldridge was a fisherman).

18. If B is a fisherman, then he doesn't own the dog; hence, he wouldn't lie and claim he did. If B is a hunter, then he does, in fact, own the dog, but he wouldn't be truthful and claim that he did. So it is not possible that B would claim to own the dog. Therefore, A lied; so A is the hunter and owns the dog.

19. If B were a hunter, he could never have made the true statement that at least one of them is a hunter. Therefore, B must be a fisherman. This means that there is at least one fisherman present (namely B); so A's statement is true; so A is also a fisherman. Also, since B is a fisherman, his statement that at least one of them is a hunter must be true. Therefore, C must be a hunter, and he is the only hunter. Since a hunter is present, the thief is present, and since the thief is a hunter, he must be C.

20. If Momba had answered, "Yes," then the judge would have had no way of knowing whether Momba was innocent or guilty (because it could be that Momba is a fisherman and the aardvark was stolen either by him or by another fisherman, and it could also be that he is a hunter, and the aardvark was stolen either by him or by another hunter). But the judge did know; therefore, Momba must have answered, "No." This means that Momba must be innocent—for the same reasons given in the solution to the first problem of this chapter (namely, that if Momba is a fisherman, then his statement is true; so a hunter stole the aardvark, and if Momba is a hunter, then his statement is false; so a fisherman stole the aardvark).

[Incidentally, it wasn't until years later that the true thief who stole the aardvark was found. It turned out to be a hunter by the name of *McSnickle*, but the trial was a dull one; so I won't bother you with it].

21. The statement made by the first twin on the first day was trivially true; hence, the first twin must be a fisherman.

Now, we consider the second day. We are not told whether A answered *yes* or *no* to the question of whether he was the second twin of yesterday, but we are told that the judge knew whether he was innocent or guilty after the answer. Now, the only way the judge could have known is by getting the answer "Yes." [Had A answered, "No," he could have been either the first twin telling the truth or the second twin lying, and the judge would have known no more than he knew before.] So A claimed to be the second twin of yesterday. Now, if he were the first twin, he could not have lied and claimed to be the second twin (because we already know that the first twin is a fisherman); so A really was the second twin and was telling the truth. Therefore, the second twin of yesterday is a fisherman. Therefore, his statement of yesterday was true; so no fisherman on this island ever steals whales. Since both brothers were fisherman, both were innocent.

22. I will first prove that E and F cannot possibly have the same occupations, and hence that G must be a hunter.

Suppose E is a fisherman; then C and D are both fishermen, and A and B are both hunters. Then Karl is a hunter and Momba is a fisherman. Then F's statement is false, since neither Karl is a fisherman, nor is Momba a hunter. This proves that if E is a fisherman, then F is a hunter.

Now suppose E is a hunter; then C and D are not both fishermen. Then A and B are not both hunters; so at least one of their statements is true. This means that either Karl is a fisherman or Momba is a hunter—just as I said! This makes F a fisherman. So if E is a hunter, F is a fisherman.

I have just proved that if E is a fisherman, then F is a hunter, and if E is a hunter, then F is a fisherman; so E and F cannot have the same occupation. This conclusively proves that G is a hunter.

Now that we know that G is a hunter, we consider H's statement. A fisherman could never claim any statement that says, among other things, that his occupation is that of G (who is a hunter); so H must be a hunter. Then the first clause of his statement is true (he is of the same occupation as G), and since his total statement is false, then it must be the second clause that is false. In other words, it is false that at least one of the defendants is innocent. So both defendants are guilty (which, incidentally, is not surprising, since it is not easy for one person, unaided, to steal a whale).

BOOK III

KING ARTHUR AND HIS DOGS

CHAPTER VII

WE GET STARTED AGAIN

It all happened at another birthday party of my young friend Alice. The party was given by her friend Arthur Louis Stephenson and his brother Robert Louis Stephenson (not the famous author, of course). Present at the party were several children, philosophers, children of philosophers, and philosophical children. Even the famous philosopher, Professor Proofsnortle, was there! [He was called "Proofsnortle" because whenever he gave a long philosophical proof, he usually would snort several times in the middle.]

First, Professor Proofsnortle addressed Alice's younger brother Tony and gave a long proof that Tony really existed. Not one of us could find a single error in his logical reasoning.

"But I already knew that," said Tony.

"Not really, snorted Proofsnortle. "You only *believed* it. Now you know it *must* be true by pure logic!"

This somehow didn't impress Tony very much.

"How about some music?" suggested Mrs. Stephenson.

"Excellent idea!" they all cried.

Well, fortunately, my friend, the eminent Italian opera singer, Maestro Baritoni, was there and held us spellbound with some arias from *Aida*.

"How about some logic puzzles?" asked Michael, one of Alice's friends.

"All right," I replied. "I have here two possible birthday presents for Alice. This one, wrapped in gold, I will call the *gold prize*, and the other one, wrapped in silver, I will call the *silver prize*. The gold prize is by far the better of the two. Now, here are the rules: Alice is to make a statement. If the statement is true, then I will give her one of the two prizes—not saying which one! But if her statement is false, then I won't give her either prize."

"And so," I said, turning to Alice, "it's up to you. Can you make a statement such that I will have no choice but to give you the gold prize?"

"I have it," said one of the group. "Alice should say, 'You will give me the gold prize.' That would do it!"

"Not necessarily," I replied; "I might be mean and say, 'You are wrong, so I won't give you either prize.' "

"But also, you might be nice and give her the gold prize," suggested Tony.

"Quite true." I replied, "I would have the choice of either giving her the gold prize or not giving her any prize at all. But the idea is that Alice should make a statement that won't leave me any choice in the matter—she should say something such that I *must* give her the gold prize."

"I have it!" suggested another. "Alice should say that she would prefer to get the gold prize. I'm sure *that* statement is true."

"So am I," I replied, "but that wouldn't commit me to giving her the gold prize. I could give her the *silver* prize for having made a true statement."

The group was silent.

"I give up," said one.

"So do I," said another.

"Now just a minute," said Alice, "Let's not give up so easily. I think I have an idea!"

Alice thought a while longer and finally came up with a statement such that I had no option but to give her the gold prize.

Problem 1 What statement would work? [The solution is given at the end of the chapter.]

"Nice work!" I said to Alice. "Now, suppose I had made the following rule instead: If you make a *false* statement, then I will give you one of the two prizes, but if you make a true statement, then I won't give you either prize. Could you then have won the gold prize?"

Problem 2 What statement would then work?

"I have a better idea," I said. "Suppose I had made the following rule: If you make a true statement, then I'll give you either the gold prize or the silver prize, or possibly both. But if you make a false statement, then you don't get either prize. Could you then have made a statement that could win you *both* prizes?"

Problem 3 What statement would win both prizes?

Postscript Alice solved this puzzle, so, in fact, I did give her both prizes.

I should like to raise a question concerning the first puzzle: Suppose Alice had said, "You won't give me either prize." What should I have done? Should I have given her a prize or not, and if so, which one?" [I'll discuss this at the end of the chapter.]

SOLUTIONS

1. Alice said: "You will *not* give me the silver prize." Now, what are my options? If I give her the silver prize, that will make Alice's statement false, but I can't give her a prize for a false statement. Therefore, I cannot give her the silver prize. Thus, Alice's statement was true, and so I must give her one of the prizes for a true statement. It cannot be the silver prize, so it must be the gold prize.

2. For this puzzle, a statement that works is: "You will give me the silver prize." The reader who understood the solution to the first puzzle should be able to prove that the above statement works.

3. This is a bit trickier. A statement that works is: "You will either give me both prizes or neither prize." Let's see why this works.

If I gave her neither prize, that would fulfill her statement and make it true, but I cannot give *no* prize if the statement is true. Therefore, I must give her at least one of the two prizes. If I gave her exactly one prize, that would make her statement false (she said, in effect, that I *won't* do that), and so I would be giving her a prize for a false statement, which I cannot do. And so I have no alternative but to give her both prizes.

Solution to the Postscript Puzzle If she had said, "You won't give me any prize," I would have been stuck; whatever I did would be wrong! For, suppose, I gave her a prize. Then her statement would be false, and so I would be giving a prize for a false statement, which is against my own rules. On the other hand, if I didn't give her a prize, then her statement would be true, and so I would have failed to give her a prize for a true statement, again violating my own rules. And so, whatever I did would be wrong. [Incidentally, this once happened at another party, and so the joke was on me!]

CHAPTER VIII

KING ARTHUR AND HIS HUNTING EXPEDITION

"You know," said Tony, "for a long time you've been promising to tell us the story of King Arthur and his dogs. How come you haven't?"

"Well," I replied, "since we are at Arthur's house, I think today would be a good time.

"However," I added, "before I tell you about King Arthur and his dogs, I would like to tell you some good old puzzles about King Arthur's hunting expedition."

The group listened attentively to the following story:

One day, King Arthur took twenty-four of his favorite knights on a hunting expedition. They stayed several days in one of King Arthur's hunting lodges in the forest. In this house there were nine rooms. The king slept in the central room, and the twenty-four knights, who were to act as his guards, were to be disposed so that there should be nine on each side of the lodge. They were placed three in a room, as in the following diagram:

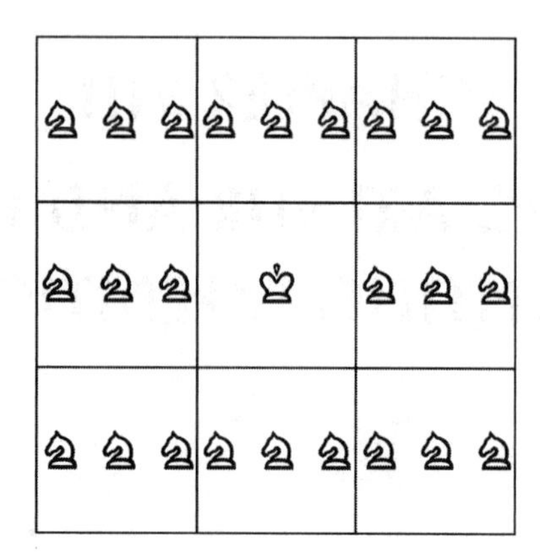

The knights asked if they might meet in the evenings in one another's rooms for games. To this King Arthur agreed, but only on condition that there should always be nine knights on each side of the lodge.

•1•

The First Night On the first night, King Arthur, before retiring to rest, made his rounds of the lodge and counted the number of knights on each side, to see that his orders were obeyed and that none of the knights had gone to the village, which was close by. He found that there were just nine on each side, and so he went to bed feeling that all was well.

But his knights had played a little prank on him! Four of them had actually sneaked away to the village. Yet the remaining knights, by a clever rearrangement, had contrived to maintain the full number of nine on each side of the lodge.

How did they do this?

•2•

The Second Night On the second night, instead of any of the knights going to the village, four of the villagers, who were their friends, came to the hunting lodge disguised as knights, which was against the rules. But when King Arthur looked around, he thought all was well, because there were nine and nine only on each side of the lodge.

How did they manage this?

•3•

The Third Night On the third night, eight visitors came, and now there were thirty-two men (other than King Arthur) in the house, but as the king still found nine on each side, he did not notice the new additions.

How was this arranged?

•4•

The Fourth Night The knights had so much fun with all of this that on the next night they received twelve visitors, not eight! Yet these thirty-six men cleverly arranged themselves to fool the king again.

How did they do this?

•5•

The Fifth Night On the fifth and last night, instead of inviting their friends to the lodge, they arranged matters so that six of them could go to the village, and there would still be nine men on each side.

How did they do this?

SOLUTIONS

1. *First Night*:

♘♘♘♘	♘	♘♘♘♘
♘	♔	♘
♘♘♘♘	♘	♘♘♘♘

2. *Second Night*:

♘♘	♘♘♘♘♘	♘♘
♘♘♘♘♘	♔	♘♘♘♘♘
♘♘	♘♘♘♘♘	♘♘

3. *Third Night*:

♘	♘♘♘♘♘♘♘	♘
♘♘♘♘♘♘♘	♔	♘♘♘♘♘♘♘
♘	♘♘♘♘♘♘♘	♘

4. *Fourth Night*:

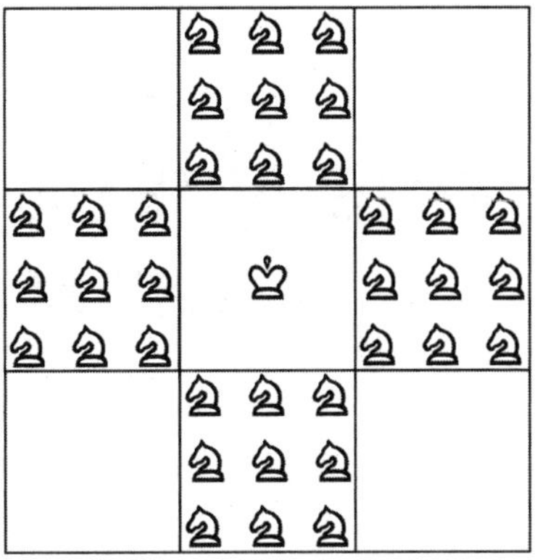

5. *Fifth Night*:

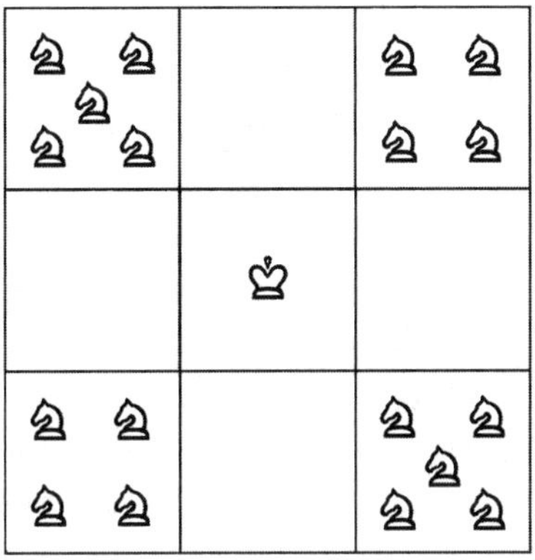

CHAPTER IX

KING ARTHUR AND HIS DOGS OF THE ROUND TABLE

"Now," I said, "I'll tell you about King Arthur's dogs of the Round Table."

"Just a minute!" interrupted Alice. "There weren't *dogs* at the Round Table; there were *knights*!"

"Which Round Table are you talking about?" I inquired.

"You mean there was more than one?" cried Alice in astonishment.

"Of course!" I replied. "There were two Round Tables—one for the knights and one for the dogs."

"How come Sir Thomas Malory never mentioned the other table?" Alice asked. "I read Malory's history of King Arthur, and he never mentioned any second Round Table!"

"Perhaps he didn't know about it," I lamely suggested.

"Then how come *you* do?" asked another.

"That's a secret!" I replied.

"I don't believe a word of what you're saying!" said Alice emphatically. "I think you're making it all up out of your own head!"

"Now, now!" I said sternly, "Do you want to hear this story or don't you?"

"We do! We do!" they all cried.

"Very well then, but no more interruptions! Now, the story is a very interesting one, and . . ."

"But is it *accurate*?" asked Alice.

"Well," I replied, "I can't guarantee that *everything* in it is *one-hundred percent* accurate in *all* respects; there might be a *few* inaccuracies every now and then."

"A few, indeed!" said Alice.

"As I was saying," I continued, ignoring the last remark, "King Arthur was sitting with his dogs at the Round Table . . ."

"How many dogs were there?" asked Tony, who is very inquisitive.

•1•

How Many Dogs Does King Arthur Have? "Good question," I replied. "Only, do you want to know how many dogs were at the Round Table, or how many dogs King Arthur had all together? Not all his dogs were privileged to be at the Round Table!"

"I'd like to know both," said Tony. "First of all, how many dogs did King Arthur have all together?"

"Well," I said, "if you add one-fourth of the number to one-third of the number, you will have ten more than one-half of the number."

How many dogs does King Arthur have?

•2•

And the Dogs of the Round Table? "What about the dogs of the Round Table; how many of them were there?" asked Tony.

"Well," I replied, "some of the dogs of the Round Table were shaggy and others were not . . ."

"How many of them were shaggy?"

"Fourteen of the dogs were shaggy," I replied. "Also, some of the dogs were male and the others were female . . ."

"You don't say!" said Alice.

"I guess that was a bit obvious, wasn't it?" I replied. "Anyway, twelve of the dogs were male. Also, some of the dogs were large and the others were small; there were thirteen large ones. The number of large shaggy dogs was four; the number of large males was three; and the number of shaggy males was five. Only one dog was large, male, and shaggy, and each dog at the Round Table was either large, male, or shaggy."

How many dogs were at the Round Table?

•3•

Queen Guinevere "What about Arthur's wife, Queen Guinevere?" asked Alice. "Did she have her own dogs?"

"No," I replied, "but she had lots of cats."

"How many?" asked Tony.

"Well," I said, "one-third of them were black, one-fourth of them were white, and the remaining ten cats were gray."

How many cats were there?

•4•

King Arthur's Collie King Arthur's collie, Alcibiades, weighed 60 pounds plus one-third of his weight.

How much did the collie weigh?

•5•

The Dog That Got Lost One of King Arthur's dogs once got lost in the mountains for five days. He walked a certain distance the first day, and on each of the other days he walked one mile more than he had on the previous day. At the end of the five days, the dog returned exhausted, since he had walked altogether 55 miles.

How many miles did he walk on the last day?

•6•

The Dogs That Ran Away One day, 59 of King Arthur's dogs ran to another town. The townspeople recognized them and took them back to Camelot in horse-drawn carts. Each large cart held nine dogs, and each small cart held four dogs.

How many large carts and how many small carts were used?

•7•

Gawaine and Uwaine Sir Gawaine and Sir Uwaine were brothers.

"Why don't we have a little joust?" said Gawaine to Uwaine one day. "And to make it interesting, why don't we agree that the loser gives the winner one hunting-dog?"

"Agreed," said Uwaine.

Now, if Gawaine loses, he will have the same number of hunting-dogs as Uwaine. If Gawaine wins, then he will have three times as many as Uwaine.

How many hunting-dogs does each brother own?

•8•

Merlin's Wolves Merlin didn't keep dogs or cats as pets, as do we ordinary folk. No! Merlin, since he was a wizard, had a pack of wolves!

Well, one afternoon, one-third of the pack ran off into the forest. Shortly after, two-thirds of the remaining wolves ran after their brothers. And soon after that, the remaining twelve wolves joined the rest in the forest. Late that night, they all came home again.

How many wolves did Merlin have?

* * *

And now, I wish to tell you two fascinating puzzles about a greyhound that is one of King Arthur's dogs of the Round Table.

•9•

The Greyhound Chases a Hare One day the greyhound leaped after a hare. When the greyhound first spotted the hare, the hare had made sixty leaps in front of the greyhound. Now, the greyhound makes two leaps in the same time as the hare makes three, but the greyhound goes as far in three leaps as the hare goes in seven.

In how many leaps did the greyhound catch the hare?

•10•

How Long Is the Greyhound? The greyhound runs past a given point in one second. At the same speed, he runs through an eighteen-foot tunnel in four seconds. How long is the greyhound?

* * * *

"Those were nice arithmetical puzzles," said Michael (after we had gone through the solutions), but I prefer logic puzzles. Do you know any more logic puzzles?"

"Oh, indeed!" I replied. "Do any of you know the story of King Arthur's search for his dog?"

"Now, just a minute!" said Alice. "King Arthur was in search of the *Holy Grail*, not his dog!"

"Oh, that was a different search altogether," I replied. "Just because King Arthur searched for the Holy Grail doesn't mean that he never searched for his dog, does it?"

The group nodded assent.

"It's really a great story," I said, "but it is rather long, and it's getting too late to tell it now."

"I have an idea," suggested Mrs. Stephenson. "Next week is Tony's birthday. Why don't we have another party here, and you can tell it then."

The group clapped their hands in delight, and even Professor Proofsnortle gave an amiable sort of snort.

And so, if you wish to know the remarkable story of King Arthur in search of his dog (which is really the main item of this volume), you are cordially invited to read the chapters that follow.

SOLUTIONS

1. It's easiest to convert everything to twelfths. Well, one-fourth is three-twelfths, and one-third is four-twelfths, so one-fourth added to one-third is seven-twelfths. Also, one-half is six-twelfths. Therefore, seven-twelfths of the number of dogs is 10 more than six-twelfths of the number, which means that one-twelfth of the number is 10; so King Arthur has 120 dogs.

2. Every dog of the Round Table is of exactly one of the following eight types:

(1) Large shaggy male;
(2) Large shaggy female;
(3) Large ordinary male;

(4) Large ordinary female;
(5) Small shaggy male;
(6) Small shaggy female;
(7) Small ordinary male;
(8) Small ordinary female.

We will have to find out how many dogs there are of each of these eight types. Then, since no dog is of more than one of these types, we add the eight numbers together, and this gives us our answer.

(1) We are given that only one dog is of this type.

(2) Since there are 4 large shaggy dogs, and just one of them is male, 3 of them are female. So there are 3 dogs of type (2).

(3) Since there are 3 large males, and just one of them is shaggy, then there are 2 large ordinary males.

(4) We have counted 6 large dogs so far, so of all the 13 large dogs, the remaining 7 must be ordinary females.

(5) Since there are 5 shaggy males, and only one of them is large, there are 4 small shaggy males.

(6) Of the 14 shaggy dogs, one is of type (1), 3 are of type (2), and 4 are of type (5); this totals 8. The rest of the shaggy dogs must be of type (6), so there are 6 dogs of type (6).

(7) Of the 12 male dogs, one is of type (1), 2 are of type (2), and 4 are of type (5); this totals 6. The remaining 5 male dogs must be of type (7).

(8) There are no dogs of this type, since we were given that every dog of the Round Table is either large, male, or shaggy.

Totaling the number of dogs of each type, there are 28 dogs of the Round Table.

3. Again, we convert to twelfths. Four-twelfths of the cats are black, and three-twelfths are white, so the remaining five-twelfths are gray. Therefore, 10 is five-twelfths the total number of cats; so the total number is 10 times twelve-fifths, which is 24.

Let us check: One-third of 24 is 8; so there are 8 black cats. One-fourth of 24 is 6; so there are 6 white cats. This means 14 cats so far, and there are 10 left over from the 24, which are gray.

4. The difference between Alcibiades's full weight and one-third of his weight is 60 pounds; but, also, this difference is two-thirds of his weight. Therefore, two-thirds of his weight is 60 pounds; one-third is 30 pounds; and his full weight is 90 pounds.

5. Fifty-five miles in five days averages out to eleven miles per day. Now, since the dog's increase was regular each day (one mile each day), then on the third day he must have hit the exact average. So he walked 11 miles on the third day. This means he walked 9 miles the first day, 10 miles the second, 11 miles the third, 12 miles the fourth, and 13 miles the fifth. These five numbers add up to 55.

6. We can solve this by subtracting 9 from 59, then 9 from the result, then 9 from that result, and so forth, until we hit a multiple of 4. So we begin: 59, 50, 41, 32. Well, 32 is divisible by 4. So the dogs can be carted in 3 large carts (making 27) and 8 small carts (making 32), which totals 59.

7. If Gawaine loses, he will lose one dog and then have half the total number of dogs between them. Therefore, Gawaine now has one more than half the total. If Gawaine wins, then he will win another dog; so he will then have two more dogs than half the total. But, also, Gawaine will have three-quarters of the dogs (since he will have three times as many as Uwaine), which is one-quarter more than one-half. So, two dogs more than one-half of the dogs is the same as one-quarter of the dogs more than one-half of the dogs. Therefore, one-quarter of the dogs is the same as two dogs. Therefore, there must be eight dogs all together. This means that Gawaine has five dogs, and Uwaine has three.

Let us check: If Gawaine loses, they will each have four dogs; if Gawaine wins, he will have six, and Uwaine will have two, and six is three times two.

8. After one-third ran off, two-thirds were left. Two-thirds of those who were left is $\frac{2}{3} \times \frac{2}{3}$ of the whole pack, and $\frac{2}{3} \times \frac{2}{3} = \frac{4}{9}$. So four-ninths of the pack went off in the second run. Also, three-ninths ran off in the first run (because three-ninths is the same as one-third). So, in the first two runs, seven-ninths of the pack went off (seven-ninths is three-ninths plus four-ninths), which left two-ninths. Therefore, twelve wolves is two-ninths of the pack; so the whole pack contains nine-halves of twelve, which is 54.

Let us check: Merlin has 54 wolves. At first, one-third of them, which is 18, ran off, leaving 36. Then, two-thirds of the 36, which is 24, ran off, leaving 12.

9. Since the greyhound makes two leaps in the same time as the hare makes in three, he makes six leaps in the same time as the hare makes nine. Also, since three hound-leaps is equal in distance to seven hare-leaps, then six hound-leaps is equal to fourteen hare-leaps. So, in every six hound-leaps, the greyhound has gone fourteen hare-leaps, and the hare has gone nine hare-leaps; so the greyhound has gained five hare leaps on the hare, which is one-twelfth of the sixty hare-leaps originally separating them. Therefore, the hound must make 72 leaps (which is 12×6) in order to catch the hare.

10. Because the greyhound runs through an eighteen-foot tunnel in four seconds, it does *not* mean that the greyhound runs eighteen feet in one second! No, to run through an eighteen-foot tunnel is to run eighteen feet *plus the dog's own length*! [If he ran just eighteen feet, he would be stuck with the end of his snout at the end of the tunnel, and the rest of the dog would still be inside.] Therefore, the dog runs eighteen feet, plus his own length, in four seconds. Now, he runs his own length in one second (because he runs past a given point in one second). Therefore, in the remaining three seconds he runs eighteen feet. So he runs eighteen feet in three seconds, which is six feet in one second. But he runs his own length in one second; so his length is six feet.

BOOK IV

THE GRAND SEARCH

CHAPTER X
MERLIN'S PLAN

Wonderful, indeed, as is the well-known story of King Arthur's search for the Holy Grail, the lesser-known tale of Arthur's search for his missing dog is at least as remarkable. As you will see, it explains the true secret of Merlin's well-deserved fame as a magician—how his marvelous skill at the art of logical deduction enabled him to solve mysteries beyond the ken of most persons of his time.

The story begins on the day that King Arthur's favorite dog was missing. All Camelot was in an uproar!

"How could he *ever* have jumped over the palace fence?" asked King Arthur in amazement. "It seems like a miracle!"

"Perhaps one of the guards carelessly let him by," suggested Merlin.

"Well, can't you use your magic and bring him back?"

"Alas, no," replied Merlin. "My magic works only on wolves, not on dogs."

"Then I guess it is hopeless for me to find my dog again," said Arthur mournfully.

"I wouldn't say *hopeless*," replied Merlin, "but, the matter will be very difficult!"

"What hope is there?" asked King Arthur eagerly.

"Well," replied Merlin, "there is Klug, the Dog Wizard."

"Klug?" asked Arthur. "I have never heard of him. What does he do?"

"Why, Klug has the power to change any dog into a wolf. If you meet Klug, all you have to do is pronounce the name of your dog. Then Klug will utter his magic incantations, and wherever in the land your dog may be, he will instantly be transformed into a wolf."

"What good will that do *me*?" cried Arthur. "I don't want my dog to be a wolf; I want him as a dog!"

"Ah!" said Merlin, "the whole point is that once your dog has become a wolf, I will have magic power over him. Then I can easily call him back to the castle!"

King Arthur thought this over for a moment.

"Still," said Arthur, "I will have him back only as a wolf, and I want him back as a dog!"

"Oh, that's no problem," replied Merlin. "Once we have him back as a wolf, I can easily change him back to his former shape. Remember, I have complete magic power over all wolves!"

"Great! Great!" cried King Arthur, clapping his hands in joy. "Now our problem is completely solved! It only remains to find Klug!"

"That *only* is more than a little," replied Merlin gravely. "Unfortunately, Klug is extremely difficult to find!"

"Where is he?" asked Arthur.

"Well you see, in recent months, Klug has become pretty much of a hermit. He has gotten extremely interested in a branch of knowledge he is developing called "Dog-logic." He is doing strenuous research in this area and does not wish to be disturbed. Hence, he has hidden himself from the world and has become a hermit living in a cave."

"Good grief!" said Arthur. "*No one* knows where he lives?"

"Yes, there is one person in the land—only one—who knows his whereabouts. His name is Gunter. Gunter is also a wizard.

If you can find Gunter, then you can find Klug. But you cannot find Klug without first finding Gunter."

"How do I find Gunter?"

"Unfortunately, Gunter is also very difficult to find," replied Merlin.

"Then the case *is* hopeless!" cried Arthur.

"No," replied Merlin, "I said that Gunter is very difficult to find; I did not say it was impossible. I wish you'd listen more carefully to what I say!"

"I am listening!" cried Arthur in despair. "Just tell me what to *do*!"

"Well, you see," replied Merlin, "one difficulty is that Gunter goes under an assumed name. He is now living in a small hamlet named Caxton. There are thirty people living in this hamlet, and five of them are wizards. One of the five wizards is Gunter.

"Now comes the difficult part," continued Merlin. "The trouble is that all of the people in Caxton, as well as in the surrounding region, are either dishonest and always lie, or are honest and always tell the truth. The dishonest ones never tell the truth, and the honest ones never lie. This makes it very difficult to evaluate the truth of anything they say!"

"What about the five wizards?" asked Arthur. "Are they honest or dishonest?"

"Some of them are honest, and some of them are dishonest," replied Merlin, "and this further complicates matters."

Arthur thought about this for a while. "Exactly what do you suggest we do?"

"We had best go immediately to Caxton," replied Merlin. "When we get there, we will try to ferret out the five wizards and bring them back to Camelot. When we have them here, I will devise further strategies to find out which one is Gunter. Then, as I have explained, once we have found Gunter, we can find Klug. Once we have Klug, we get him to turn your dog into a wolf. Then I use my magic power over the wolf to bring him back to the castle, and then I turn him back into a dog."

This sounded like an excellent plan, so after making a few hasty preparations, the two set out on their journey. Little did they realize what strange adventures awaited them!

CHAPTER XI

TWO UNEXPECTED OBSTACLES

Towards late afternoon, Arthur and Merlin found themselves in a rather heavily wooded region.

"We are not far from Caxton," said Merlin, "and have already entered the region where everyone is either honest and always tells the truth, or is dishonest and always lies. We must now be very careful!"

At this point, they came to a fork in the road, but the sign to Caxton had been blown down by a storm.

"Zounds!" said Merlin. "I don't remember whether the left road or the right road is the one that leads to Caxton!"

Just then, they spied two inhabitants standing under a tree.

"Let us ask them," suggested Arthur. "They probably know."

"I have no doubt that they *know*," replied Merlin. "My only doubt is about their honesty!"

Well, they approached the two inhabitants. "Which road leads to Caxton?" asked Arthur.

"The left one," said the first inhabitant.

"Is he honest?" Arthur asked the second inhabitant.

"No, he is not," was the reply.

"At least one of us is honest," said the first inhabitant, with a strange smile.

At this point, Merlin knew which road to take. Did they take the left road or the right road? The solution follows.

* * *

"How did you know which road to take?" asked Arthur, after they were well on their way.

"Oh, that was easy" replied Merlin. "That was about the simplest case I ever came across. If only our remaining adventures turn out to be that simple!"

"You still haven't told me," remarked Arthur.

"Oh," replied Merlin, "I hardly thought it needed explanation. The second one said that the first one was dishonest. This means that one of the two is honest and the other is dishonest, because if the second one is honest, his claim is true, which means that the first one is dishonest. On the other hand, if the second one is dishonest, then he lied about the first one, which means that the first one is honest. Therefore, the two are of opposite types—one is honest and the other is dishonest."

"But which is the honest one?" asked Arthur.

"I wouldn't have known," replied Merlin, "if the first one hadn't made a second statement, but fortunately he said that at least one of the two was honest, and what he said was true. Since he told the truth, he is honest; therefore, he also told the truth when he said that the left road leads to Caxton. That is why we took the left road."

* * *

Well, the next obstacle proved more difficult. Arthur and Merlin soon came across another fork in the road, and again the sign to Caxton had been blown down by the storm.

"Zounds!" said Arthur. "Again we are foiled!"

"Not necessarily," replied Merlin thoughtfully. "Do you not spy yonder three damsels resting under the tree?"

"In sooth, I do," replied Arthur, "and fairer damsels have I rarely spied, but are they honest?"

"They are the three Cornwall sisters," replied Merlin. "Their names are Helen, Lynette, and Vivian. I have heard many wondrous tales about them, but I do not know which of them are honest and which of them are dishonest."

Arthur and Merlin soon came up to the three sisters. "Which road leads to Caxton?" asked Arthur.

Helen then named either the *left* road or the *right* road, but she said it too indistinctly for either Arthur or Merlin to make out what she said.

"What did she say?" Merlin asked Lynette.

"She said the *left* road," replied Lynette.

"She did not!" said Vivian.

Arthur did not know what to make of this. Then Merlin asked, "Exactly how many of you three are honest?"

Helen then either said *one*, or she said *two*, or she said *three*, but she said it too softly for either Arthur or Merlin to make out.

"What did you say?" Merlin asked Lynette.

"She said *one*," Lynette replied.

"Did she answer truthfully?" Merlin asked Lynette.

"Yes, she did," Lynette replied.

At this, Merlin stood for a while in thought. "Come," he said finally to Arthur, "I know which road we should take"

Which road did they take? The solution follows.

* * *

"This seems much more difficult than our last problem," Arthur said to Merlin when they were well on the road. "How did you ever solve this one?"

"To begin with," said Merlin, "Lynette and Vivian made contradictory claims; Lynette said that Helen said 'left,' and

Vivian said that Helen didn't. Therefore, one of these two lied, and the other told the truth. So one of them—Lynette or Vivian—is honest, and the other dishonest."

"All well and good," said Arthur, "but how do we know which of them is the honest one?"

"By considering Lynette's second statement," replied Merlin. "Lynette said that Helen said that exactly one of the three is honest. Mind you, Lynette never said that exactly one of the three is honest; she only said that Helen *said* that. Now, could Helen really have said that?"

"Why not?" asked Arthur.

"Let's look at it this way," replied Merlin. "Since exactly one of them—Lynette or Vivian—is honest, then the number of honest ones among the three depends entirely on whether or not Helen is honest. If Helen is honest, then two of the three are honest, and if Helen is dishonest, then only one of the three is honest. Right?"

"Of course," replied Arthur, "but how does that help us?"

"If Helen were honest" Merlin continued, "then exactly two of the three would be honest, but then Helen, since she was honest, couldn't have lied and said that exactly one is honest; she would have told the truth and claimed that two were honest."

"That's clever!" said Arthur.

"On the other hand," continued Merlin, "if Helen were dishonest, then there would, in fact, be exactly one honest one among them; hence, Helen wouldn't have told the truth and said *one*; she would have lied and either said *two* or *three*."

"I see!" said Arthur.

"Therefore," continued Merlin, "whether Helen is honest or dishonest, she couldn't possibly have said *one*, so Lynette lied when she said that Helen said *one*. This proves that Lynette must be dishonest."

"Very clever," said Arthur, "but I still don't see how that tells us which is the correct road."

"For that," replied Merlin, "we must take into account Lynette's final statement: Lynette said that Helen told the truth—in other words, Lynette is claiming that Helen is honest. Since Lynette is dishonest, her claim is false, so Helen is not really honest; Helen is dishonest, just like Lynette. Also, Lynette claimed that Helen said that the left road leads to Caxton, but since Lynette lied, Helen never said that; Helen must have really said that the right road leads to Caxton. Since Helen said that the right road leads to Caxton, and Helen is dishonest, then it is really the left road that leads to Caxton. That is why I chose the left road."

"Marvelous!" cried Arthur in genuine admiration. "I see that your reputation is well deserved! How did you ever learn to reason like that?"

"It's just a question of simple logic," Merlin replied. "Almost anyone with a little practice can learn the art of logical reasoning. I daresay that by the time our present adventures are over and we have recovered your dog, you yourself will be quite adept at this kind of reasoning."

"Do you *really* think we will find my dog again?" asked Arthur, with a trace of sadness in his voice.

"We can only try," replied Merlin.

CHAPTER XII
THE SEARCH GETS UNDERWAY

Arthur and Merlin arrived in Caxton without any further difficulties.

"Let us put up at this inn," suggested Merlin. "Tomorrow you go forth and interview the inhabitants of this hamlet. Any wizard you come across, tell him he must return with us to Camelot. Any non-wizard you come across—just forget about him and pass on to the next inhabitant. Whenever you can't decide whether someone is or is not a wizard, return to me immediately and give me an exact report of the interview. Perhaps we can then reason it out. Meanwhile, I have certain preparations of my own that I must make here."

THE STORY OF MADOR

Early the next morning, Arthur set out to interview the inhabitants. The first one he met was named *Mador*.

"Just what are you?" asked King Arthur.

"I am a dishonest wizard" was the curious reply.

Well, King Arthur did not know what to make of this. Can it be determined whether or not Mador is a wizard? Also, was Mador telling the truth? The solution follows.

* * *

King Arthur went right back to the inn to consult Merlin.

"I don't know what to make of this," Arthur confided. "Mador claimed to be a dishonest wizard, but perhaps he was lying. What is he really?"

"Did you summon him to Camelot?" asked Merlin.

"No," replied Arthur. "I had no idea what to do."

"Just as well you didn't," said Merlin. "Mador is no wizard, though he certainly is dishonest!"

"How do you know that?" asked Arthur.

"It's obvious!" replied Merlin. "To begin with, could Mador possibly be honest?"

Arthur thought about this for a bit.

"I guess not," replied Arthur, "but how can we tell whether or not he is a wizard?"

"Just a minute," said Merlin, who wished to test Arthur's understanding, "you say that Mador is not honest; what reasons do you have to believe that?"

"If he were honest, said Arthur, "then what he said would be true—which would mean that he is a dishonest wizard. But one who is honest can't be a dishonest wizard; he can't be a dishonest anything!"

"Very good!" said Merlin. "Now that we know that Mador is dishonest, can't you tell whether or not he is a wizard?"

"How?" asked Arthur.

"By considering his statement," replied Merlin. "He claimed to be a dishonest wizard, but since he lied, he is *not* a dishonest wizard. So he is dishonest, but not a dishonest wizard. Therefore, he is not a wizard at all!"

"I think I see," said Arthur, "but I'm not absolutely sure. Can you go over that last point again?"

"If he were a wizard," explained Merlin, "what kind of wizard could he be? We've just seen that he is not a dishonest wizard, and he can't be an honest wizard either, since he is not honest. Therefore, he can't be a wizard."

"Ah, now I see!" said Arthur.

THE STORY OF BONDCOINE

The next inhabitant whom Arthur interviewed was named *Bondcoine.*

"Tell me something about yourself," said Arthur. "Are you a wizard? Are you honest?

"I am not an honest wizard" was the reply.

Arthur was puzzled by this. Was this really the same situation as the last one?

Can it be determined whether Bondcoine is a wizard? Arthur had to go back and consult Merlin. The solution follows.

* * *

"No, it's not the same case," said Merlin, after Arthur had told him of the interview. "Like Mador, Bondcoine is no wizard, but unlike Mador, Bondcoine is honest."

"Please explain why," asked Arthur.

"To begin with, a dishonest person can't be an honest wizard, can he?"

"Of course not!" replied Arthur. "A dishonest person can't be honest, period."

"Exactly," said Merlin. "Now, if Bondcoine were dishonest, then his statement that he is not an honest wizard would be true, but the dishonest ones around here don't make true statements; therefore, he can't be dishonest; he must be honest."

"I follow you so far," replied Arthur.

"Well, then," continued Merlin, "now that we know that Bondcoine is honest . . ."

"Wait!" interrupted Arthur. "I think I see the rest myself. Since Bondcoine is honest, then his statement that he is not an honest wizard is true. So he is not an honest wizard, yet he is honest. Therefore, he can't be a wizard. Is my reasoning right?"

"Absolutely!" said Merlin. "I told you that with a little practice you will start getting the knack of it!"

THE STORY OF MODRED

"This logic is really an amazing thing" thought Arthur, as he walked forth from the inn. "I wonder who ever invented it!"

Well, the next inhabitant Arthur met was named *Modred*.

"Are you an honest wizard?" asked Arthur.

"I am neither honest nor a wizard," replied Modred.

What can be deduced about Modred? The solution follows.

* * *

"Very curious," said Arthur to Merlin. "Modred claimed to be neither honest nor a wizard. Can you tell anything about him?"

"Praise be to Heaven!" exclaimed Merlin, "We have found our first wizard! He is not an honest one, but he is a wizard—and that's what really counts. When you see him next, be sure to summon him to Camelot."

"How do you know he is a wizard?" asked Arthur.

"Well, now," replied Merlin, "I think you should be able to solve this yourself. The first question to ask yourself is whether Modred is honest. Is he?"

Arthur thought for a moment. "Oh," he said, "Modred must be dishonest—no honest person would claim that he is neither honest nor a wizard!"

"Very good," said Merlin. "Now that you know that he is dishonest, can't you deduce the rest?"

"Well," said Arthur, "let me see: Since Modred is dishonest, his claim was false. This means that it is *not* the case that he is neither honest nor a wizard. Now, what does it mean that it is false that he is neither honest nor a wizard? This is a bit confusing. Ah! I see it; it means that either he is honest or he is a wizard. Right?"

"Right, so far," replied Merlin.

"So," continued Arthur, thinking aloud, "either he is honest or he is a wizard. However, he is not honest, as we have seen; therefore, he must be a wizard."

"Better and better!" said Merlin. "You are making excellent progress."

THE STORY OF GRIFLET

The next inhabitant Arthur met was named *Griflet*. Somehow, he looked to Arthur as if he might be a dishonest wizard.

"Are you, by any chance, a dishonest wizard?" Arthur asked.

"I am at least one or the other," Griflet replied. "Either I am dishonest or I am a wizard."

What can be deduced about Griflet? The solution follows.

* * *

"I believe I can solve this one myself," Arthur said to Merlin. "According to my reasoning, Griflet is a wizard—but an honest one, not a dishonest one."

"What are your reasons?" asked Merlin.

"Well," replied Arthur, "to begin with, Griflet can't be dishonest, for if he were, then he really would be one or the other (dishonest or a wizard) as he said, but a dishonest person here doesn't make true statements. Therefore, it is impossible that he is dishonest; he must be honest."

"Right, so far," said Merlin.

"Since he is honest," continued Arthur, "his statement is true, which means that he is either dishonest or a wizard. But he is not dishonest, so he is a wizard."

"Excellent!" said Merlin. "Your analysis was perfect."

THE STORY OF ONTZLAKE AND MARKHAUS

The next adventure is the most interesting one yet.

Merlin decided to go with Arthur for the next interview. They came across two inhabitants named *Ontzlake* and *Markhaus*.

"Tell us something about yourselves," said Merlin.

"We are alike as far as our honesty goes," said Ontzlake. "Either we are both honest or we are both dishonest."

Merlin and Arthur stood a while in thought. Then one of the two inhabitants, Ontzlake or Markhaus, pointed to the other one and said, "He is a wizard, but I am not." Merlin was then able to deduce whether or not Ontzlake was a wizard and whether or not Markhaus was a wizard.

Now, I have not told you whether it was Ontzlake or Markhaus who made the second statement—the reason is that the various historical documents I have consulted are not clear on this point. Nevertheless, I have given you enough information to settle the matter.

Is Ontzlake a wizard or not? Is Markhaus a wizard or not? If you find yourself baffled at this point, the solution follows.

* * *

SOLUTION

From the first statement, which was definitely made by Ontzlake, it follows that Markhaus must be honest, although the honesty of Ontzlake cannot be determined. Here is the reason why Markhaus must be honest:

If Ontzlake is honest, his statement was true, which means that he and Markhaus are really alike; hence, Markhaus must also be honest. This proves that if Ontzlake is honest, so is Markhaus. Now, what if Ontzlake is dishonest? Well, in that case, his statement was false, which means that he and Markhaus are *not* alike. In other words, Markhaus is not dishonest like Ontzlake. So in this case, Markhaus is again honest. This proves that regardless of whether Ontzlake is honest or dishonest, Markhaus must be honest.

Now, we are definitely given the fact that Merlin *did* solve the problem. (If we were not given that, then there would be no way for *us* to solve the problem.) The next question is: Which of the two made the second statement? Well, if it was Ontzlake, then Merlin couldn't possibly have solved the problem, because he had no way of knowing whether Ontzlake was honest or not. Therefore, it must have been Markhaus who said: "He is a wizard, but I am not," and Merlin, knowing that Markhaus was honest (for the reasons I have already given), knew that his statement was true. In other words, Ontzlake is a wizard, but Markhaus is not.

So the answer is that Ontzlake is a wizard, but Markhaus is not. However, we don't know whether Ontzlake is an honest wizard or a dishonest one.

CHAPTER XIII

THE DIFFICULTIES DOUBLE

"Well," said Merlin to Arthur that evening, "our first day has been quite successful. We have found three of the five wizards, Modred, Griflet, and Ontzlake. Let us hope that tomorrow we will find the other two without difficulties."

The next morning, Arthur went out alone to interview the remaining inhabitants. (Merlin had to remain behind to make more of his own preparations.) As it turned out, the cases to be solved were about twice as difficult as those encountered on the previous day. Fortunately, Arthur received help from Bondcoine.

Bondcoine, we recall, turned out to be an honest inhabitant of the hamlet, though he is not a wizard. King Arthur met him early in the morning and confided his mission to him.

"Merlin and I know that there are five wizards in this hamlet," explained Arthur, "and we have already found three of them—Modred, Griflet, and Ontzlake. It remains for us to find the other two."

"That is interesting!" said Bondcoine, "I am not surprised about Modred and Griflet, though I had no idea that Ontzlake was a wizard! Is he an honest or a dishonest wizard?"

"We do not yet know," replied Arthur, "nor does it yet matter. The important thing is to find the other two wizards. Do you know who they are?"

"Alas no," replied Bondcoine. "For one thing, the wizards here tend to be rather secretive about their wizardry. For another, I have resided here only a short time and don't yet know the inhabitants too well. But whatever information I do have is fully at your service."

Well, at this point I shall tell of Arthur's interviews that day, assisted by Bondcoine. Unlike the last chapter, I shall not incorporate the solutions into the story, but shall give all the solutions at the end of the chapter.

MICHAEL AND BELVIDERE

"Here come *Michael* and *Belvidere*," said Bondcoine. "I don't know too much about them, but you might try questioning them."

Well, Michael claimed that neither of them was a wizard. Belvidere claimed that neither of them was honest. Is either of them a wizard, and if so, which one?

ALISANDER AND PALIMEDES

"Then there are *Alisander* and *Palimedes* over there," said Bondcoine. "Perhaps one or both of them are wizards. Why not try them?"

Well, Alisander claimed that at least one of the two was a wizard. Palimedes claimed that at least one of the two was dishonest.

What can be deduced about them?

GARLON AND GORLIAS

"Here come *Garlon* and *Gorlias*; they are brothers," said Bondcoine.

Arthur asked the two brothers to tell him something about themselves.

"One of us is honest, and one of us is not," said Gorlias.

"My brother is not a wizard," said Garlon.

"Are you a wizard?" Arthur asked Garlon.

"Yes, I am," Garlon replied.

What can be deduced about Garlon and Gorlias?

ERBIN AND EDEYON

"Ah, here I might be of some help," said Bondcoine. "I know those two brothers over there; their names are *Erbin* and *Edeyon*. Everyone here knows that one of them is honest, and the other one is dishonest, but no one seems to know which brother is the honest one."

Well, Erbin claimed that his brother was an honest wizard, and Edeyon claimed that his brother wasn't a wizard at all.

Which, if any of them, are wizards?

ACCOLON, TURQUINE, AND TOR

"I know those three," said Bondcoine. "Their names are *Accolon*, *Turquine*, and *Tor*. I know for sure that Accolon is not a wizard, but I don't know about Turquine or Tor."

"Well, Arthur approached the three and said, "Tell me something about yourselves."

"All three of us are dishonest," said Accolon.

"That's not true; only one of us is dishonest," said Turquine.

Then Tor pointed to one of the other two and said, "He is a wizard."

Later, when King Arthur recounted this interview to Merlin, he had forgotten whether Tor had pointed to Accolon or

Turquine. Nevertheless, Merlin was clever enough to figure out whether or not Tor was a wizard, and whether or not Turquine was a wizard. (Accolon, we recall is not a wizard.)

Is Tor a wizard? Is Turquine a wizard?

* * *

MERLIN'S SOLUTIONS

Merlin was able to solve all these cases, and here are the solutions in Merlin's own words.

Michael and Belvidere If Belvidere were honest, he would never have lied and claimed that neither he nor Michael was honest. Therefore, Belvidere must be dishonest. Since he is dishonest, his statement that neither of them is honest was false, which means that, in reality, at least one of them is honest. Since it is not Belvidere who is honest, it must be Michael. This proves that Michael is honest. It then follows that Michaels's statement was true, so neither of them is a wizard.

Alisander and Palimedes If Palimedes were dishonest, then it would be true that at least one of the two is dishonest, but a dishonest one cannot make true statements. Therefore, it is impossible that Palimedes is dishonest. Since he is honest, his statement was true that at least one of them is dishonest. Since it is not Palimedes who is dishonest, it must be Alisander. Since Alisander is dishonest, his statement that at least one of them is a wizard is false. Therefore, neither of them is a wizard.

Garlon and Gorlias It first must be proved that Garlon is dishonest. Well, Gorlias is either honest, or he isn't. Suppose he is; then his statement that he and Garlon are of different types is true, which means that Garlon is dishonest. So if Gorlias is honest, then Garlon is dishonest. If, on the other hand, Gorlias is dishonest, then his statement that he and Garlon are of different types is false, which means that Garlon is really of the same type as Gorlias, hence, also dishonest. So, whether Gorlias is honest or dishonest, Garlon is dishonest, in either case.

Now that we know that Garlon is dishonest, we know that both of his statements were false; hence, Gorlias is a wizard, but Garlon is not.

Erbin and Edeyon To begin with, could Erbin be honest? If he were, we would get the following contradiction: If Erbin is honest, his claim is true, which means that Edeyon is an honest wizard, hence, honest; hence, Erbin and Edeyon are both honest. But this is contrary to what Bondcoine said; therefore, it cannot be true that Erbin is honest.

Now that we know that Erbin is dishonest, it follows that his statement was false; hence, Edeyon is not an honest wizard. Yet Edeyon is honest (since he is opposite to Erbin), so he is not a wizard. Also, since Edeyon is honest, his statement about Erbin was true. Therefore, Erbin is not a wizard. This proves that neither one is a wizard.

Accolon, Turquine, and Tor To begin with, if Accolon were honest, he wouldn't have lied and said that all three are dishonest. Therefore, Accolon is dishonest. This means that he lied when he said that all three are dishonest. Therefore, at least one of them really is honest; it must be Turquine or Tor (and possibly both). It cannot be determined whether Turquine is honest or dishonest, but, in either case, Tor must be honest for the following reasons: We know that at least one of the two, Turquine or Tor, is honest. If Turquine is dishonest, then it must be Tor who is honest. But what if Turquine is honest? Then his statement that only one is dishonest must be true—which means that Accolon is the only honest one; hence, Tor must, again, be honest. So in either case (whether Turquine is honest or not), Tor is honest.

Now that we know that Tor is honest, we see that he couldn't have pointed to Accolon (who isn't wizard) and said that he is a wizard. Therefore, Tor must have pointed to Turquine, and since Tor is truthful, Turquine is a wizard.

Now that we know that Turquine is a wizard, what about Tor? Well, since all five wizards have now been found (Turquine being the last), Tor cannot be a wizard.

In summary, Accolon is neither honest nor a wizard. Tor is honest, but he is not a wizard. As for Turquine, he is a wizard, but we don't yet know whether he is honest or dishonest.

CHAPTER XIV
THE GRAND TRIAL

Now comes the most remarkable part of the entire adventure.

All five wizards have now been found. Their names are Modred, Griflet, Ontzlake, Gorlias, and Turquine. Of course, one of them is really Gunter under an assumed name, but we have no idea which one. At any rate, the net seems to be closing in!

All five wizards are taken back to Camelot for the great trial to find out which one was really Gunter. What follows is the most famous interrogation in the entire history of English wizardry!

The five wizards were taken to the trial room. Merlin conducted the investigation.

"I know that one of you is really Gunter," said Merlin to the five defendants, "and I intend to find out which!"

"Gorlias is not Gunter," said Ontzlake.

"Turquine is not Gunter," said Gorlias.

"Modred is not Gunter," said Griflet.

"Griflet is Gunter," said Modred.

At this point Merlin spoke. "I do not yet know which of you is Gunter, but I now know two of you who are definitely *not* Gunter. *You* are not Gunter," he said, pointing to one of the five, "and *you* are not Gunter," he said, pointing to another, "so you two are free to leave the court."

The two happily left the court and went back to Caxton. This left three wizards still on trial.

"*You* haven't said anything," said Merlin to Turquine (who was one of the remaining three). "What do you have to say about all this?"

"Well," replied Turquine, "the three of us here are not all of the same type; at least one of us is honest, and at least one of us is dishonest."

Merlin thought about this for some time.

"I still don't know which of you is Gunter," he said, "but I now know another one who is not Gunter." Merlin then pointed to one of the three and said, "You are not Gunter, so you may leave the court."

This defendant left the court, which left two wizards on trial. Merlin then asked one of them, "Are you two of the same type?"

"No, we are not," was the reply. "One of us is honest and the other is not."

At this point, you have all the information needed to determine which of the five wizards is Gunter. Is it Modred, Griflet, Ontzlake, Gorlias, or Turquine? The solution follows.

* * *

SOLUTION

To solve this, we must recall that Modred is dishonest, and Griflet is honest. Since Griflet is honest, his statement was true, so Modred is not Gunter. Since Modred is dishonest, his statement is false, so Griflet is not Gunter. Therefore, it was Modred and Griflet who were first dismissed from the court. This left Ontzlake, Gorlias, and Turquine.

Now, Ontzlake said that Gorlias is not Gunter, and Gorlias said that Turquine is not Gunter. They couldn't have both been lying, because if Ontzlake were lying, then Gorlias is Gunter, and if Gorlias were lying, then Turquine is Gunter. But Gorlias and Turquine can't both be Gunter; hence, either Ontzlake or Gorlias must be telling the truth. This proves that Ontzlake and Gorlias are not both dishonest.

We next prove that Turquine must be honest. Well, suppose Turquine were dishonest; then his statement would be false, which means that all three *would* be of the same type (since he claims they are not), which would mean that all three must be dishonest. However, we have seen that the other two can't both be dishonest. This proves that Turquine can't be dishonest, so Turquine must be honest.

Since Turquine is honest, his statement is true; so at least one of the three must be dishonest. If Ontzlake is dishonest, then Gorlias is Gunter. If Gorlias is dishonest, then Turquine is Gunter. In neither case is Ontzlake Gunter. Therefore, it has not been determined that Ontzlake is not Gunter, but so far either of the other two could be. So it must have been Ontzlake who was next told by Merlin that he may leave the court.

This left Gorlias and Turquine. Now, Merlin asked one of them (we are not told which) whether he was of the same type as the other, and he answered that he was not. Since Turquine is honest, then it is impossible that Gorlias could claim to be of a different type than Turquine, because if Gorlias were honest, then he would truthfully say that he is of the same type as Turquine, and if Gorlias were dishonest, he would falsely say that he is of the same type as Turquine. Therefore, it was Turquine, not Gorlias, to whom Merlin asked the question. So Turquine claimed that he and Gorlias are not of the same type, and since Turquine is truthful, then they are really not of the same type, which means that Gorlias is dishonest. Therefore, Gorlias's statement was false, which means that Turquine is Gunter.

CHAPTER XV

BUT WHERE IS THE DOG?

"So, Gunter, we have found you at last!" exclaimed Arthur joyfully. "So *you* are Gunter the wizard, and it turns out that you are an honest wizard at that; you always tell the truth."

"Of course, I always tell the truth," replied Gunter. "What sense is there in lying?"

"I'm glad you feel that way," replied Arthur. "Unfortunately, not everybody does. Anyway, I certainly had enormous trouble finding you! Do you know why we need you?"

"I haven't the slightest idea," replied Gunter.

"Well," replied Arthur, "you know Klug, the dog wizard?"

"Of course," replied Gunter.

"Klug is actually the one we are after," said King Arthur, "but Merlin tells me that you are the only wizard in the land who knows where Klug is."

Gunter thought for a moment. "Is it quite correct to say that I am the *only* wizard who knows where Klug is?" asked Gunter. "After all, Klug is also a wizard, and surely he knows where he is, doesn't he?"

"All right," said Arthur angrily. "Technically you are right, but that's the trouble with you pedantic logicians! You really know what I meant!"

"I insist you talk to me purely rationally!" said Gunter.

"Very well," said Arthur, "I'll correct myself. What I *should* have said is that you are one of the only two wizards in the land who knows where Klug is—the other one is Klug himself."

"Then you would be wrong!" said Gunter. "It so happens you were right the first time."

"But that's impossible!" cried Arthur. "You yourself just proved that: Since Klug is also a wizard, then he, too, knows where he is."

"I have it!" said Merlin. "The only possibility is that Klug is no longer in the land. Is that it?"

"No, no," replied Gunter, "I believe Klug is somewhere in this land."

"I have it!" said Arthur. "Klug is probably now asleep, so he doesn't know where he is. Is that it?"

"No," replied Gunter, "I have no reason to believe that Klug is now asleep."

Both Arthur and Merlin thought this over for a very long time.

"I give up!" said Merlin.

"I, also!" said King Arthur.

"You gentlemen are not very clever!" said Gunter. "It obviously follows from what I have told you that I must be Klug."

"What!" cried Arthur in amazement.

"Of course, it follows: Since I know where Klug is, and Klug knows where Klug is, and I am the *only* one who knows where Klug is, then how could I be anyone other than Klug?"

"*You* are really Klug!" gasped Arthur, still unable to believe his ears.

"None other," replied Klug, with a bow.

"But I thought you were Gunter," said Merlin.

"Gunter was my first assumed name," replied Klug, "but after it became known that Gunter knew where Klug was, I thought I had better assume another name and thus remain undisturbed. And so, as you know, I then assumed the name of Turquine."

"But I thought you were living in a cave," cried Arthur. "How is it that I found you living in the hamlet of Caxton?"

"At first I lived in a cave," replied Klug, "but then I reasoned that since it was known that I had become a hermit and that hermits usually live in caves, people would search all the caves in the land to ferret me out. Therefore, I left the cave, assumed the new name of Turquine, and retired to the quiet hamlet where you found me."

"Wonderful!" cried Arthur. "So at last we have Klug himself! Now, Klug, you know why I need you?"

"You probably want to hear all about my latest research in canine logic," replied Klug. "Well, my latest result is particularly interesting."

Klug then went on to explain all about his work in this area. His account was highly theoretical and completely beyond the comprehension of Arthur, who had neither the logical nor the mathematical background to follow.

"I have no idea what you are talking about and couldn't care less!" shouted Arthur angrily. "Right now, I have no time for all this theoretical nonsense!"

"Nonsense?" said Klug, paling with rage. "You dare to call my research *nonsense*? That does it! I'll have nothing to do with you again!"

So saying, Klug started packing his things, preparing to depart.

"Now, just a minute, Klug!" said Arthur. "I haven't gone to all this trouble finding you for nothing! I have lost a dog, and I want you to turn him into a wolf!"

Klug continued packing, totally ignoring what Arthur had said.

"Now look here, Klug," said Arthur, getting angrier by the minute. "Do you realize that if you disobey me, I can have you hanged for insubordination?"

"Oh, really now?" said Klug with a smile, as he slowly surely disappeared into thin air.

* * *

"Zounds!" said Merlin. "That foils all our hopes. We'll never be able to find him again!"

Poor Arthur! After all that trouble in finding the five wizards and then finding the one who was Klug—alias Gunter, alias Turquine—after all that trouble, *just* when the plan was about to succeed, Arthur spoiled it by tactlessly hurting Klug's feelings.

Well, what can Arthur do now? Nothing, he decided. Sadly, he left the throne room, went upstairs to his private chamber, threw himself on the bed, and wept. He had so loved that dog!

As he was weeping, he heard a whining sound, and his dog came out from under the bed, where he evidently had been hiding all the while.

I shall not go into the joyful details of the reunion. I leave it to your imagination to realize how happy King Arthur was to find his dog again. I might only add that the dog was equally happy.

Why was the dog hiding under the bed all this time? This is one mystery that no historian has yet successfully unraveled. One theory is that dogs sometimes get into a strangely temperamental mood for several days at a time, and, in this condition, hide from everybody. [Incidentally, I can personally testify that one of my dogs has had two such spells in her life.]

Another theory is that the dog was not hiding under the bed at all, but had really escaped the palace grounds, and that it was really Klug who had brought him back. According to this theory, Klug, despite his impertinence and gruff exterior, was really very kind at heart and could not remain angry with anyone for very long. So right after Klug disappeared, he started feeling sorry for King Arthur, and knowing that Arthur had lost his beloved dog, Klug used instantaneous magic, put the dog into a deep sleep, and transported him to the spot under King Arthur's bed where Arthur would happily find him after he retired.

Those historians who favor the second theory offer two facts in its support. For one thing, several days after the dog was found, Arthur did claim that his dog appeared unusually sleepy when he emerged from under the bed. Therefore, it is well possible that the dog had been in a deep sleep. The second fact is that several months after the entire incident, there were rumors that Klug had been discussing the matter with some fellow wizards and told them that he considered Merlin's plan for recovering the dog to be unnecessarily roundabout. As Klug was reported to have said: "It was not necessary to first turn the dog into a wolf and then have *Merlin* bring him back; I'm perfectly capable of doing the job myself!"

Now, these alleged facts—particularly the second—must be taken with a grain of salt, because, for one thing, the rumors are open to question. And even if the rumor about Klug is correct, Klug was known to be quite competitive towards Merlin, and may simply have been boasting. However, why would an honest wizard like Klug boast? Therefore, it may well be that the first theory is correct after all—the dog may have been hiding under the bed the whole while.

So, as matters now stand, neither theory has been fully substantiated. It is my hope that one day a graduate student, working for a Doctor's degree in English or history, may do some research in this area and bring some decisive facts to light.

At any rate, this is all that I know about this mysterious matter, and so, I conclude my tale.

EPILOGUE

The last time during that summer that I got together with the children was the Sunday before they went back to school.

"This has been a wonderful summer," said Alice. "We all thoroughly enjoyed it, and we learned so much!"

"Yeah," said Tony, "it's much better than going to school!"

"Where did you ever get all those wonderful puzzles?" asked Alice.

"Well," I replied, "the puzzles were mainly of two types—arithmetical and logical. The arithmetic puzzles are based on ideas of puzzles that are very old; indeed, I dug them up from some pretty well-forgotten sources. It's a shame they haven't been around for a while, since they are both entertaining and instructive. Of course, I tailored many of them to suit the particular occasion."

"Some of these arithmetic puzzles are like those in algebra courses," remarked William.

"I know," I replied. "The main novelty in the way I showed them to you was in the methods of solution, rather than the puzzles themselves."

"What do you mean?" asked William.

"Well, you see, many of these puzzles are of the same type as I got from these old children's books. Now, as you said, many of them can be routinely solved using algebra, but these books were written for kids who didn't yet know any algebra. And so these books gave these lovely commonsense logical solutions that I showed you.

"In fact," I continued, "when I first came across these puzzles, I tried solving them *without* using algebra, but I couldn't.

I was, in a way, spoiled by knowing algebra! Only after going through half a dozen or so of their solutions, did I get the hang of the tricks and was then able to do the rest without algebra."

"Do you think it is better to know these logical commonsense methods than to know algebra?" asked William.

"I am not saying that," I replied. "I believe one should know both. But I firmly believe that these logical commonsense methods should be learned *before* one learns algebra. Otherwise, algebra is apt to become a purely mechanical routine—often used by students as a substitute for genuine thinking!"

"That's true," said William. "I was extremely interested in comparing your types of solutions with those taught to us in algebra. One shouldn't say that either is better than the other; they should both be understood. Only, I agree that your type of reasoning should be learned first."

"What about the logic puzzles?" asked Alice. "Where did you get those?"

"Oh, those I made up," I replied. "Logic is my special hobby."

"You know, Raymond," said Michael, "I think that you should put all these together and make them into a book, so other kids can have fun with them."

"Good idea," I replied, "except that I've forgotten most of them by now."

"Oh, I've got them all carefully written down in a notebook," said Alice.

"Me, too," said Charlie. "I figured you might need them again, and I know how forgetful you absent-minded professors are!"

Some time later, I went through Alice's and Charlie's notebooks and was amazed at how much thought and care they had bestowed on these problems! They wrote down not only the solutions I have given in this book, but also many alternative solutions of their own. They also considered many interesting side issues. All this only confirmed my belief that if children are not *forced* to learn a subject, but approach it fully in the spirit of a hobby, they learn so much more!

Well, let us be thankful to all these children for making this book possible. Next summer, I'll probably get together with them again, and if any interesting new puzzles or stories come up, I promise to let you know about it.

About the Author

Curiously enough, I have lived four different lives—as a mathematician, musician, magician, and author of essays and puzzle books. I was born in 1919 in Far Rockaway, New York. As a child, I was equally interested in music and science. In high school I fell in love with mathematics, and was torn between becoming a mathematician or a concert pianist. My first teaching position was at Roosevelt College in Chicago, where I taught piano. At about that time I unfortunately developed tendonitis in my right arm forcing me to abandon piano performances as my primary career. As a result of this I turned my attention to mathematics. I had learned most of this on my own, with very little formal education at the time. I then took a few advanced courses at the University of Chicago, and supported myself at the time as a professional magician!

Strangely enough, before I had a college degree, or even a high school diploma, I received an appointment as a mathematics instructor at Dartmouth College on the basis of some papers I had written on mathematical logic. After teaching at Dartmouth, the University of Chicago gave me a Bachelor of Arts degree, based partly on courses I had never taken, but had successfully taught, such as calculus. I then went to Princeton University for my Ph.D. in mathematics in 1959. I subsequently taught at Princeton, NYU, Belfor Graduate School, Lehman College and Graduate Center, and my last teaching position was as a distinguished ranks professor at Indiana University. I have published forty research papers in mathematical logic and twenty-two books. In 2009 I am coming out with four more books.

Despite the trouble with my right arm, I have been able to give concerts, and am still musically active. I am a working member of the Piano Society. I continue to write puzzle books, but these are more than mere puzzle books—it is through recreational logic puzzles that I introduce the general reader to deep results in mathematics and logic!

RAYMOND M. SMULLYAN